Synthesis Lectures on Communications

Series Editor

William H. Tranter, Virginia Tech, Blacksburg, VA, USA

This series of short books cover a wide array of topics, current issues, and advances in key areas of wireless, optical, and wired communications. The series also focuses on fundamentals and tutorial surveys to enhance an understanding of communication theory and applications for engineers.

Jerry D. Gibson

Analog Communications

Introduction to Communication Systems

 Springer

Jerry D. Gibson
Department of Electrical and Computer
Engineering
University of California
Santa Barbara, CA, USA

ISSN 1932-1244   ISSN 1932-1708   (electronic)
Synthesis Lectures on Communications
ISBN 978-3-031-19586-0   ISBN 978-3-031-19584-6   (eBook)
https://doi.org/10.1007/978-3-031-19584-6

This Springer imprint is published by the registered company Springer Nature Switzerland AG
The registered company address is: Gewerbestrasse 11, 6330 Cham, Switzerland

Contents

About the Author

Jerry D. Gibson is Professor of Electrical and Computer Engineering at the University of California, Santa Barbara. He is co-author of the books *Digital Compression for Multimedia* (Morgan-Kaufmann, 1998) and *Introduction to Nonparametric Detection with Applications* (Academic Press, 1975 and IEEE Press, 1995) and author of the textbook, *Principles of Digital and Analog Communications* (Prentice-Hall, 2nd ed., 1993). He is Editor-in-Chief of *The Mobile Communications Handbook* (CRC Press, 3rd ed., 2012), Editor-in-Chief of *The Communications Handbook* (CRC Press, 2nd ed., 2002), and Editor of the book, *Multimedia Communications: Directions and Innovations* (Academic Press, 2000). His most recent books are *Rate Distortion Bounds for Voice and Video* (Coauthor with Jing Hu, NOW Publishers, 2014) and *Information Theory and Rate Distortion Theory for Communications and Compression* (Morgan-Claypool, 2014).

Dr. Gibson was Associate Editor for Speech Processing for the *IEEE Transactions on Communications* from 1981 to 1985 and Associate Editor for Communications for the *IEEE Transactions on Information Theory* from 1988 to 1991. He was an IEEE Communications Society Distinguished Lecturer from 2007 to 2008.

In 1990, Dr. Gibson received The Fredrick Emmons Terman Award from the American Society for Engineering Education, and in 1992, he was elected Fellow of the IEEE. He was the recipient of the 1993 IEEE Signal Processing Society Senior Paper Award for the Speech Processing area. He received the *IEEE Transactions on Multimedia* Best Paper Award in 2010 and the IEEE Technical Committee on Wireless Communications Recognition Award for contributions in the area of Wireless Communications Systems and Networks in 2009.

Amplitude Modulation

1

1.1 Introduction

A common lay definition of the verb *to modulate* is "to tune to a certain key or pitch." In amplitude modulation, this is precisely what is being done; namely, the information to be transmitted, hereafter called the *message signal,* is moved to another location in the frequency spectrum. This relocation is easily accomplished by multiplying the message signal by a pure sinusoid; however, one may wonder why we wish to modulate the message signal at all. A little thought produces several excellent reasons.

First, the original frequency content of many signals that we wish to transmit overlap. For example, speech signals fall in the range 0–4000 Hz, music contains frequencies in the range 0–20,000 Hz, and television video signals originally occupy the band 0–5 MHz. If several of these signals were transmitted simultaneously over the same medium (cable, wire pairs, etc.) without modulation, they would interfere with each other and unintelligible signals would be received. Modulation allows us to send these signals simultaneously without interference by moving each signal to a different frequency band during transmission. A second reason for relocating message signals in frequency is that there are numerous sources of noiselike signals at low frequencies, such as car ignitions, electric lights, and electric motors. A third reason relates to antenna size for free-space transmission. For the efficient radiation of electromagnetic waves, the antenna size should be proportional to or larger than the transmitted wavelength. If the frequency to be transmitted is 5000 Hz, the wavelength is 60,000 meters! Of course, an antenna anywhere near this size is impractical. There are additional reasons for using various types of modulation, and they will become evident as the development progresses.

The discussion of amplitude modulation (AM) begins with a treatment of AM doublesideband suppressed carrier (AMDSB-SC) systems in Sect. 1.2. The development includes

© The Author(s), under exclusive license to Springer Nature Switzerland AG 2023
J. D. Gibson, *Analog Communications*, Synthesis Lectures on Communications,
https://doi.org/10.1007/978-3-031-19584-6_1

sketches of a typical time-domain AMDSB-SC waveform as well as its frequency content, a discussion of coherent demodulation, and a derivation of the effects of frequency and phase errors in coherent demodulation. Conventional AM (i.e., AMDSB-transmitted carrier) is presented in Sect. 1.3, which includes envelope detection, sketches of the time-domain waveform and its Fourier transform, and a discussion of percent modulation and efficiency. In Sects. 1.4 and 1.5 we develop single-sideband and vestigial sideband AM, respectively. Time-domain waveform expressions and their Fourier transforms, demodulation of the received signals, frequency and phase errors in coherent demodulation, and the selection of filter shape are addressed. Superheterodyne systems are discussed in Sect. 1.6, which includes such topics as the choice of intermediate-frequency (IF) filter center frequency and bandwidth, image stations and radio-frequency (RF) filter bandwidth, and tuning range. AM radio examples are given. Quadrature amplitude modulation is defined in Sects. 1.7 and 1.8 we introduce the concept of frequency-division multiplexing and how it is used in communication systems. Binary and multilevel amplitude shift keying (ASK) are considered in Sect. 1.9, with sketches of time-domain waveforms and a discussion of demodulation methods.

1.2   Double-Sideband Suppressed Carrier

From Fourier transform properties we know that one way of relocating a signal in the frequency domain is to multiply its time-domain waveform by a pure sinusoid. More specifically, let the message signal be denoted by $m(t)$ and let the multiplying sinusoid or *carrier signal* be $A_c \cos \omega_c t$. Forming the product of these two signals yields

$$s_{\text{DSB}}(t) = A_c m(t) \cos \omega_c t. \tag{1.2.1}$$

The frequency content of $s_{\text{DSB}}(t)$ is given by

$$S_{\text{DSB}}(\omega) \triangleq \mathcal{F}\{s_{\text{DSB}}(t)\} = \frac{A_c}{2}[M(\omega + \omega_c) + M(\omega - \omega_c)], \tag{1.2.2}$$

where $\mathcal{F}\{m(t)\} = M(\omega)$. The relocation of $m(t)$ in the frequency domain is easily seen by sketching Eq. (1.2.2). If $|M(\omega)|$ is as shown in Fig. 1.1a, then $|S_{\text{DSB}}(\omega)|$ has the form illustrated in Fig. 1.1b. For simplicity, the phases are assumed to be zero and are not shown. Nonzero phases do not alter the development.

The waveform in Eq. (1.2.1) is called an AM double-sideband, suppressed carrier (AMDSB-SC) wave, or more commonly, just a double-sideband wave. The justification for this name comes directly from Fig. 1.1b. The frequency content in the range $\omega_c \leq |\omega| \leq \omega_c + \omega_m$ is called the upper sideband (USB), and the frequency content in the range $\omega_c - \omega_m \leq |\omega| < \omega_c$ is called the lower sideband (LSB). Since physical signals are defined only for positive frequencies, $m(t)$ is totally described by the shape of the frequency content in Fig. 1.1a in the range $0 \leq \omega \leq \omega_m$. Concentrating only on positive

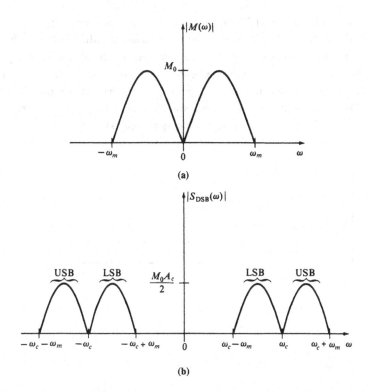

Fig. 1.1 Frequency content of the message and double sideband signals: **a** Message signal; **b** double sideband signal

frequencies, note that the USB in Fig. 1.1b has the same shape as $|M(\omega)|$ and that the LSB has the same shape except that it is rotated about $\omega = \omega_c$. Because this shape is preserved, each sideband alone has enough information to reconstruct $m(t)$. Since we have two sidebands, we use the term *double sideband*. The terminology *suppressed carrier* comes from the fact that although $\cos \omega_c t$ appears explicitly in Eq. (1.2.1), there is no frequency content in $S_{DSB}(\omega)$ at $\omega = \omega_c$ that was not originally at $\omega = 0$ in $M(\omega)$. In the following section, where a different type of AM is discussed, a carrier term will be clearly evident in the modulated waveform frequency content.

In sketching $S_{DSB}(\omega)$, it has been implicitly assumed that $\omega_c > \omega_m$. If this is not the case, then $M(\omega - \omega_c)$ and $M(\omega + \omega_c)$ will overlap, thus causing distortion. For most free-space transmission applications, $\omega_c \gg \omega_m$ ($\omega_c > 10\omega_m$).

Although the multiplication in Eq. (1.2.1) has in fact relocated $M(\omega)$ in the frequency domain, the multiplication has also doubled the required bandwidth. Again considering only positive frequencies, the bandwidth occupied by $M(\omega)$ is ω_m, but the frequency band occupied by $S_{DSB}(\omega)$ is $2\omega_m$. Since bandwidth is at a premium in many applications, this fact is important for design trade-offs with other modulation methods.

A sketch of $s_{DSB}(t)$ in Eq. (1.2.1) is shown in Fig. 1.2b for the assumed time-domain message signal sketched in Fig. 1.2a. Notice that the dashed outline in Fig. 1.2b is simply $m(t)$ and $m(t)$ flipped about the time axis. This dashed line is, of course, not physically observable, but the outline is shown in the figure, since it is commonly called the *envelope* of the waveform. Notice that in sketching Fig. 1.2b, we have again used the fact that $\omega_c > \omega_m$. This is evident since the inner waveform has frequency ω_c and oscillates several times while the amplitude of $m(t)$ is changing only slightly.

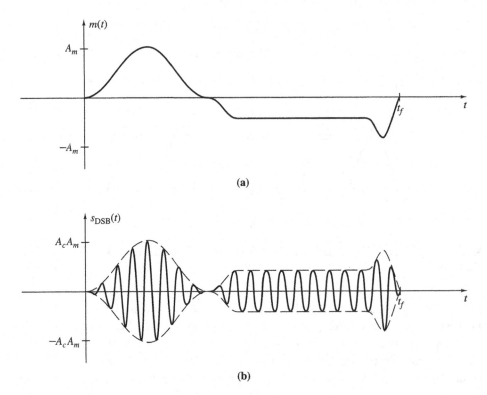

Fig. 1.2 Time-domain waveforms for Eq. (1.2.1): **a** Message signal; **b** product of $m(t)$ and $A_c \cos \omega_c t$

The signal $s_{DSB}(t)$ is the modulated waveform and would be the transmitted signal in a communication system that uses AMDSB-SC modulation. However, we have yet to demonstrate that $m(t)$ can be recovered from $s_{DSB}(t)$ and certainly this is a requirement for any useful system. To illustrate the recovery of $m(t)$, consider Fig. 1.3. Figure 1.3a represents the transmitter, while (b) is a block diagram of the receiver. Following Fig. 1.3b, we form

$$x(t) = s_{DSB}(t) \cos \omega_c t = A_c m(t) \left[\frac{1}{2} + \frac{1}{2} \cos 2\omega_c t \right]. \tag{1.2.3}$$

From Eq. (1.2.3) we see that in the frequency domain we have a scaled version of $M(\omega)$ at $\omega = 0$ and at $\omega = \pm 2\omega_c$. Thus if the low-pass filter in Fig. 1.3b is distortionless for $0 \le |\omega| \le \omega_m$ and has a cutoff frequency (ω_{co}) such that $\omega_m < \omega_{co} < 2\omega_c - \omega_m$, then $m(t)$ can be recovered exactly (within a scale factor).

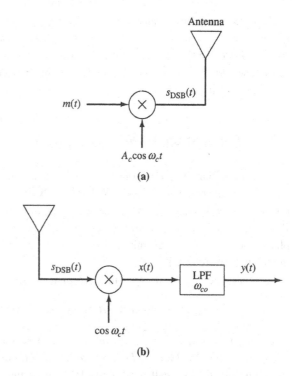

Fig. 1.3 AMDSB-SC System block diagram: **a** Transmitter; **b** receiver

The carrier signal at the receiver, $\cos \omega_c t$, is given the special name *local oscillator* (LO). The accuracy of the LO is extremely important if $m(t)$ is to be recovered undistorted. For instance, in Fig. 1.3 we assumed that the carrier used at the transmitter and the LO used at the receiver had exactly the same frequency and phase. We therefore say that the carrier and the LO are synchronized in frequency and phase, and as a result, this type of demodulation is sometimes called *synchronous* or *coherent* demodulation. If the LO is not synchronized with the carrier, distortion will result. To illustrate this claim, let the transmitted signal be $s_{DSB}(t)$ in Eq. (1.2.1) but let the LO signal be $\cos[(\omega_c + \Delta\omega_c)t + \theta]$, where we have included a frequency error $\Delta\omega_c$ and a phase error θ. Then

$$s_{DSB}(t) \cos[(\omega_c + \Delta\omega_c)t + \theta] = A_c m(t) \cos \omega_c t \cos[(\omega_c + \Delta\omega_c)t + \theta]$$

$$= \frac{A_c}{2} m(t) \{\cos[\Delta\omega_c t + \theta]$$

$$+ \cos[(2\omega_c + \Delta\omega_c)t + \theta]\}. \qquad (1.2.4)$$

If this signal is applied to the low-pass filter in Fig. 1.3b, the distorted output will be (assuming that $\Delta\omega_c$ is not too large)

$$y_d(t) = \frac{A_c}{2}m(t)\cos[\Delta\omega_c t + \theta]. \tag{1.2.5}$$

The frequency error causes $m(t)$ to be modulated, which can cause a wide variety of distortions depending on the relationship between ω_m and $\Delta\omega_c$. If we assume that $\Delta\omega_c = 0$, we see that the phase error causes $m(t)$ to be attenuated. As θ gets larger and approaches 90°, $y_d(t)$ approaches zero! Certainly, inaccurate LO frequency and phase are to be avoided if possible.

1.3   Conventional AM

The most familiar application of amplitude modulation is to AM radio. In this application, we have many receivers (radios) and relatively few transmitters (radio stations). As a result, it is highly desirable that the receivers be as simple and hence as cheap as possible. If we use AMDSB-SC, we must design the receiver such that the LO frequency and phase track that of the transmitter. Since this can be expensive, another type of amplitude modulation is preferred.

For this type of AM, the transmitted signal is given by

$$s_{AM}(t) = [A + aA_c m(t)]\cos\omega_c t = A[1 + am_n(t)]\cos\omega_c t, \tag{1.3.1}$$

where $m_n(t) = m(t)/|m(t)|_{\max} = m(t)/A_m$ is a normalized version of $m(t)$, $A = A_c A_m$, and $0 \le a \le 1$ is called the *modulation index*. We require that $A > aA_c|m(t)|_{\max}$. The signal $s_{AM}(t)$ in Eq. (1.3.1) is called AMDSB-TC (transmitted carrier), *conventional AM*, or simply AM. The reasons for the designation AMDSB-TC are most evident from the Fourier transform of $s_{AM}(t)$; however, it is clear by inspection of Eq. (1.3.1) that in addition to the product of $m(t)$ and $\cos\omega_c t$ there is another term which consists of only $\cos\omega_c t$ (the carrier). The signal in Eq. (1.3.1) is called *conventional AM*, or just AM, because of its widespread use in commercial AM radio.

If we assume that $\mathcal{F}\{m(t)\} = M(\omega)$ is as shown in Fig. 1.1a, the Fourier transform of $s_{AM}(t)$ is shown in Fig. 1.4 and is given by

$$S_{AM}(\omega) = \pi A[\delta(\omega + \omega_c) + \delta(\omega - \omega_c)] + \frac{aA_c}{2}[M(\omega + \omega_c) + M(\omega - \omega_c)]. \tag{1.3.2}$$

It is clear both from Fig. 1.4 and from Eq. (1.3.2) that there is now a separate carrier component not present in $S_{DSB}(\omega)$ [compare Fig. 1.1b and Eq. (1.2.2)]. Of course, the two sidebands remain.

Although we have established that there is, in fact, a separate carrier term in $s_{AM}(t)$, the question is: Why is it there? To address this question, let us assume that $m(t)$ is as

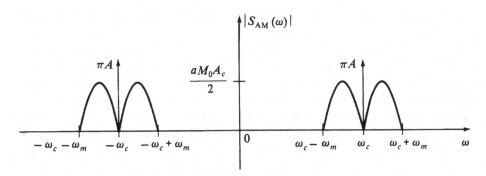

Fig. 1.4 Frequency content of conventional AM

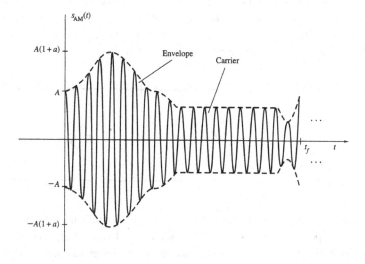

Fig. 1.5 Time-domain sketch of $s_{AM}(t)$

shown in Fig. 1.2a and sketch $s_{AM}(t)$ in the time domain. This waveform is illustrated by Fig. 1.5. The key point of interest in this figure is that the positive outer envelope of $s_{AM}(t)$ has the exact shape of $m(t)$. As a result, if we could construct a receiver that simply follows this envelope, $m(t)$ can be recovered. That this can be achieved will be discussed shortly. Continuing our discussion of Fig. 1.5, we note that we have made use of the requirement that $A > aA_c|m(t)|_{max} = aA_cA_m$. The purpose of this requirement is to guarantee that the positive envelope never goes negative, and hence that the positive envelope has the exact shape of $m(t)$. This could just as well have been guaranteed by making A larger than the modulation index times the amplitude of the most negative swing of $A_cm(t)$. However, since most waveforms of interest, for example, speech and music, have approximately equal positive and negative voltage amplitude ranges, our requirement on A is logical.

We have now established that the purpose of the extra carrier term in $s_{AM}(t)$ is to preserve the positive envelope of the transmitted signal in the shape of $m(t)$. The way we take advantage of this fact is to use an *envelope detector* at the receiver. A circuit diagram of an envelope detector is shown in Fig. 1.6. The principle behind this circuit is extremely simple. During a positive cycle of the AM wave, the diode is forward biased and the capacitor charges up to the peak value, which is the envelope. During a negative cycle of $s_{AM}(t)$, the diode is back biased and the capacitor discharges through the resistor. The principal possible source of distortion is the time constant of the discharge during negative cycles. If the capacitor discharges too rapidly, the output voltage will fall far below the envelope before the next positive cycle. This situation will cause excessive ripple at the detector output. If $\omega_c \gg \omega_m$, however, the ripples will be slight. Another possible source of distortion is called "failure to follow," and it occurs when the capacitor discharges too slowly to follow a rapid decrease in the envelope. Of course, by a judicious choice of the RC time constant these problems can be avoided.

Fig. 1.6 Envelope detector

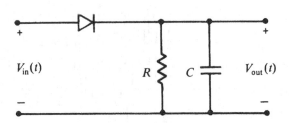

The output of the envelope detector is thus $A + aA_c m(t)$, and since it is not difficult to remove the dc component A, we can easily obtain a scaled version of $m(t)$. Note that the envelope detector requires no accurate knowledge of the carrier frequency or phase, and hence the requirement for an accurate LO signal as in AMDSB-SC has been eliminated.

It is common for conventional AM systems to be discussed in terms of "percent modulation." Such terminology relates to the modulation index, a, in Eq. (1.3.1). If $a = 1$, it is called 100% modulation, and if $a = 0.25$, it is called 25% modulation. The reason for this nomenclature is that it gives an indication of the relative maximum amplitudes of the carrier term alone and the message-carrying term in Eq. (1.3.1). If $a = 1$, then the two terms have equal amplitudes, since $A = A_c A_m$. If $a = 0.25$, the message carrying term has only one-fourth the amplitude of the free carrier term. It is intuitive that for maximum system efficiency, as much energy as possible should be allocated to the information-bearing term. For 100% modulation, this is exactly what is occurring.

1.4 Single Sideband

While both AMDSB-SC and conventional AM are successful methods for relocating a message signal in frequency, it is clear that both techniques do so by doubling the bandwidth over that required by the baseband message signal. Since the spectral shape of the message is preserved in both the upper and lower sidebands, only one of these sidebands is necessary to represent exactly the message at the receiver. Therefore, to conserve bandwidth, we now investigate modulation methods that require half the frequency allocation of the DSB techniques discussed in Sects. 1.2 and 1.3.

Modulation methods that transmit only the upper or lower sideband of the DSB signal are called *single sideband* (SSB). There are two common methods for generating SSB signals. One method consists of generating a DSB-SC signal and then filtering out the unwanted sideband. This approach requires an excellent filter and the filtering is usually performed at a lower frequency and then translated to a higher frequency. The filtering approach is most attractive for message signals that have little low-frequency content. Since frequencies below 100–200 Hz are not critical to the intelligibility of voice signals, speech can be transmitted successfully using this filtering technique for SSB generation.

The second method for generating SSB signals is the phase shift method. This approach is best introduced by an example.

Example 1.4.1 Let the message signal be a single tone, $m(t) = \cos \omega_m t$. The Fourier transform of $m(t)$ is illustrated in Fig. 1.7a. Multiplying $m(t)$ by the carrier wave, we get $s_{DSB}(t) = m(t) \cos \omega_c t$, which has the spectrum shown in Fig. 1.7b. If we pass $s_{DSB}(t)$ through a filter that retains only the lower sideband, the remaining spectral content is as shown in Fig. 1.7c. In the time domain the LSB signal is $s_{LSB}(t) = \cos(\omega_c - \omega_m)t = \cos \omega_m t \cos \omega_c t + \sin \omega_m t \sin \omega_c t$. Thus we can produce $s_{LSB}(t)$ by generating $\cos \omega_m t \cos \omega_c t$ and $\sin \omega_c t \sin \omega_m t$ and adding them together. The first term is just the product of the message and carrier signals, while the latter can be written as $\sin \omega_m t \sin \omega_c t = \cos(\omega_m t - \pi/2) \cdot \cos(\omega_c t - \pi/2)$, and hence can be obtained by separately shifting the phases of the message and carrier waves by $-90°$ and then forming their product.

Although Example 1.4.1 exhibits only a single special case, the result holds in general for all message signals, since any signal can be expressed as a sum of sinusoids over a given finite interval. The general result is that for a message signal $m(t)$, the SSB signal is

$$s_{SSB}(t) = m(t) \cos \omega_c t \pm m_h(t) \sin \omega_c t, \qquad (1.4.1)$$

where $m_h(t)$ is obtained by shifting the phase of each frequency component of $m(t)$ by $-90°$. In Eq. (1.4.1), the positive sign is associated with the LSB, while the USB is gotten with the negative sign. The component $m_h(t)$ is called the *Hilbert transform* of $m(t)$ and is given by

$$\mathcal{H}\{m(t)\} = m_h(t) = \frac{1}{\pi} \int_{-\infty}^{\infty} \frac{m(\tau)}{t - \tau} d\tau. \tag{1.4.2}$$

Since $m_h(t)$ is obtained by shifting the phase of each frequency component of $m(t)$ by $-90°$, we know that

$$\mathcal{F}\{m_h(t)\} = M_h(\omega) = \begin{cases} -jM(\omega), & \omega \geq 0 \\ jM(\omega), & \omega < 0. \end{cases} \tag{1.4.3}$$

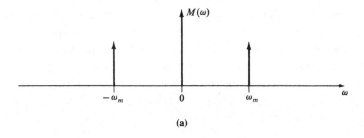

(a)

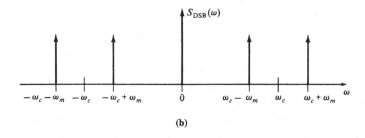

(b)

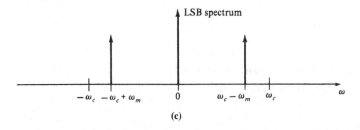

(c)

Fig. 1.7 Spectra for Example 1.4.1: **a** Message signal; **b** double sideband signal; **c** lower sideband

Using Eq. (1.4.3), we can generalize Example 1.4.1 and demonstrate that Eq. (1.4.1) represents a single-sideband signal for a general $m(t)$.[1] Consider the Fourier transform of a general message signal $M(\omega) = \mathcal{F}\{m(t)\}$, which is sketched in Fig. 1.8a. Defining

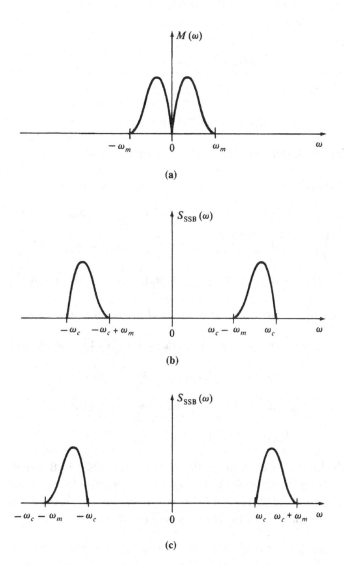

Fig. 1.8 Single sideband spectra for a general message: **a** Message spectrum; **b** spectrum of a lower sideband SSB signal; **c** spectrum of an upper sideband SSB signal

---

[1] This development is due to J. L. LoCicero (private correspondence, 1984). A similar treatment appears in Lathi (1968).

$$M_p(\omega) = M(\omega), \quad \omega \geq 0, \tag{1.4.4}$$

and

$$M_n(\omega) = M(\omega), \quad \omega < 0, \tag{1.4.5}$$

we can write

$$M(\omega) = M_p(\omega) + M_n(\omega), \tag{1.4.6}$$

and from Eq. (1.4.3)

$$M_h(\omega) = -jM_p(\omega) + jM_n(\omega). \tag{1.4.7}$$

Taking the Fourier transform of Eq. (1.4.1), we have

$$
\begin{aligned}
S_{SSB}(\omega) &= \mathcal{F}\{s_{SSB}(t)\} \\
&= \frac{1}{2}\{M_p(\omega + \omega_c) + M_n(\omega + \omega_c) + M_p(\omega - \omega_c) + M_n(\omega - \omega_c)\} \\
&\quad \pm \frac{1}{2j}\{-j[-M_p(\omega + \omega_c) + M_n(\omega + \omega_c) + M_p(\omega - \omega_c) - M_n(\omega - \omega_c)]\} \\
&= \frac{1}{2}\{M_p(\omega + \omega_c) + M_n(\omega + \omega_c) + M_p(\omega - \omega_c) + M_n(\omega - \omega_c)\} \\
&\quad \pm \frac{1}{2}\{M_p(\omega + \omega_c) - M_n(\omega + \omega_c) - M_p(\omega - \omega_c) + M_n(\omega - \omega_c)\}. \tag{1.4.8}
\end{aligned}
$$

For the positive sign on the second set of braces in Eq. (1.4.8), we obtain

$$S_{SSB}(\omega) = M_p(\omega + \omega_c) + M_n(\omega - \omega_c), \tag{1.4.9}$$

which is sketched in Fig. 1.8b. Similarly, we have for the negative sign,

$$S_{SSB}(\omega) = M_p(\omega - \omega_c) + M_n(\omega + \omega_c), \tag{1.4.10}$$

which is sketched in Fig. 1.8c. Clearly, Eq. (1.4.9) represents a LSB signal, while Eq. (1.4.10) is an USB signal, and thus Eq. (1.4.1) is valid for any $m(t)$, as claimed.

To recover $m(t)$ from the SSB signal in Eq. (1.4.1), we can use coherent demodulation, which consists of multiplying by the carrier waveform to get

$$
\begin{aligned}
s_{SSB}(t)\cos\omega_c t &= m(t)\cos^2\omega_c t \pm m_h(t)\sin\omega_c t \cos\omega_c t \\
&= \frac{m(t)}{2} + \frac{1}{2}[m(t)\cos 2\omega_c t \pm m_h(t)\sin 2\omega_c t]. \tag{1.4.11}
\end{aligned}
$$

The $2\omega_c$ components can be removed from Eq. (1.4.11) by simple low-pass filtering, and thus $m(t)$ is recovered. Of course, as in AMDSB-SC demodulation, phase or frequency inaccuracies in the LO will result in distortion of the recovered message.

The SSB signal discussed thus far is a suppressed carrier signal in that no separate carrier term is available in the transmitted signal. Having illustrated the utility of envelope detection in Sect. 1.3, we find that it is of interest to investigate the possibility of using envelope detection in conjunction with SSB transmission. Such a combination would conserve bandwidth and still be simple to demodulate. To begin, consider the (LSB) SSB-TC (transmitted carrier) signal

$$s_{\text{SSBTC}}(t) = A \cos \omega_c t + m(t) \cos \omega_c t + m_h(t) \sin \omega_c t, \tag{1.4.12}$$

which can be rewritten in the form

$$s_{\text{SSBTC}}(t) = \left\{ [A + m(t)]^2 + m_h^2(t) \right\}^{1/2} \cos[\omega_c t + \theta], \tag{1.4.13}$$

where

$$\theta = -\tan^{-1}\left\{ \frac{m_h(t)}{A + m(t)} \right\}. \tag{1.4.14}$$

If the SSB-TC is applied to an envelope detector, the envelope detector output will be

$$\begin{aligned}
\rho(t) &= \left\{ A^2 + m^2(t) + 2Am(t) + m_h^2(t) \right\}^{1/2} \\
&= A\left\{ 1 + \frac{m^2(t)}{A^2} + \frac{2m(t)}{A} + \frac{m_h^2(t)}{A^2} \right\}^{1/2}.
\end{aligned} \tag{1.4.15}$$

If $A \gg |m(t)|$ and $A \gg |m_h(t)|$, the envelope is approximately

$$\rho(t) \cong A\left\{ 1 + \frac{2m(t)}{A} \right\}^{1/2}. \tag{1.4.16}$$

Using a series expansion for the square root and neglecting terms of second order and higher, we obtain

$$\rho(t) \cong A + m(t). \tag{1.4.17}$$

Removing the dc term (A), we are left with $m(t)$, as desired.

Unfortunately, to arrive at Eq. (1.4.17) we had to assume that $A \gg m(t)$ and $A \gg m_h(t)$. Although this can usually be achieved in practice (but not always), it is extremely inefficient to allocate so much power to the carrier term. As a result, envelope detection is not nearly as important in SSB transmission as it is for conventional AM.

1.5   Vestigial Sideband

A single-sideband signal can be generated by first producing an AMDSB-SC signal and then filtering out one of the sidebands, or the phase shift method demonstrated in Example 1.4.1 can be used. The former method is most successful on messages that have little low-frequency content and hence do not require exceptionally sharp filter cutoff characteristics. The phase shift method can be employed for messages with frequency content down to dc, but the design of a 90° phase shift network over a wide range of frequencies is not an easy task. Therefore, in those instances where SSB generation is difficult, we must search for another modulation method, which is implementable but still bandwidth efficient.

Vestigial sideband (VSB) AM transmission has both of these characteristics. VSB relaxes the stringent sharp cutoff requirements of SSB by retaining a vestige or trace of the unwanted sideband in the transmitted signal. As long as only a portion of the "extra" sideband is sent, VSB occupies less bandwidth than DSB; however, VSB always requires greater bandwidth than SSB. Vestigial sideband signals are generated by approximately filtering a DSB-SC waveform.

Fig. 1.9 Vestigial sideband system: **a** Transmitter; **b** receiver

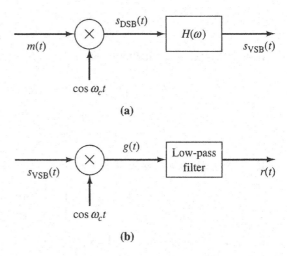

Of course, the filter characteristics for obtaining a VSB signal from a DSB signal cannot be chosen arbitrarily.

The constraints on the filter transfer functions for VSB can be established by considering the communication system block diagram shown in Fig. 1.9. In this diagram, $H(\omega)$ produces a VSB signal from the just-generated DSB signal, and we desire to determine any requirements on $H(\omega)$ such that $m(t)$ can be recovered by synchronous or coherent demodulation, as indicated in Fig. 1.9b. By inspection of Fig. 1.9a, we have that $\mathcal{F}\{s_{\mathrm{VSB}}(t)\} = S_{\mathrm{VSB}}(\omega)$ is given by

$$S_{VSB}(\omega) = \frac{1}{2}[M(\omega + \omega_c) + M(\omega - \omega_c)]H(\omega), \qquad (1.5.1)$$

where $\mathcal{F}\{m(t)\} = M(\omega)$. At the receiver we have $g(t) = s_{VSB}(t)\cos\omega_c t$, so

$$G(\omega) = \frac{1}{4}\{[M(\omega + 2\omega_c) + M(\omega)]H(\omega + \omega_c)$$
$$+ [M(\omega) + M(\omega - 2\omega_c)]H(\omega - \omega_c)\}. \qquad (1.5.2)$$

The receiver low-pass filter removes the components at $\pm 2\omega_c$, so that the Fourier transform of $r(t)$ is

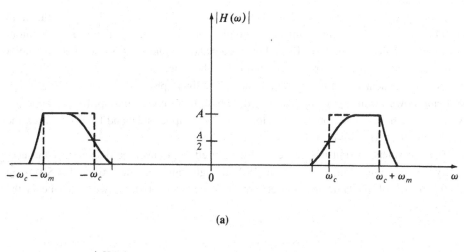

(a)

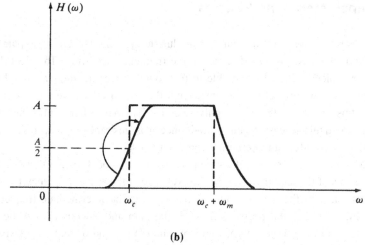

(b)

Fig. 1.10 VSB filter magnitude requirement: **a** Magnitude of $H(\omega)$; **b** complementary symmetry

$$R(\omega) = \frac{1}{4}M(\omega)[H(\omega + \omega_c) + H(\omega - \omega_c)]. \qquad (1.5.3)$$

For distortionless reception of $m(t)$, we require that

$$R(\omega) = CM(\omega)e^{-j\omega t_d}, \quad |\omega| \leq \omega_m, \qquad (1.5.4)$$

where C is a constant, ω_m is the highest frequency present in $m(t)$, and t_d is a time delay. Directly from Eq. (1.5.4), we find that

$$H(\omega + \omega_c) + H(\omega - \omega_c) = Ce^{-j\omega t_d} \qquad (1.5.5)$$

for $|\omega| \leq \omega_m$.

If we ignore the phase of $H(\omega)$ [assume that $\angle H(\omega) = 0$], the requirement in Eq. (1.5.5) necessitates that $H(\omega)$ have complementary symmetry about ω_c, as demonstrated graphically in Fig. 1.10. If the lower-sideband response is pivoted about the point $H(\omega_c)$ and an ideal filter response is obtained in the range $\omega_c \leq \omega \leq \omega_c + \omega_m$, the filter has the complementary symmetry requirement. The linear phase characteristic is less easy to attain; however, in many applications, the phase requirement may prove unimportant. Although we have carried out the development for an upper-sideband bandpass filter, the results are of course valid for the lower-sideband case.

Vestigial sideband transmission can also be used with envelope detection if the necessary carrier term is injected. As in SSB, though, envelope detection of VSB is an inefficient use of available transmitted power, since most of the power must be in the carrier signal.

1.6   Superheterodyne Systems

Broadcast AM radio uses conventional AM modulation (AMDSB-TC). Each pair of adjacent radio stations is spaced 10 kHz apart in the frequency spectrum. To select the radio station of our choice, we need to be able to tune our receiver in frequency, and we also need a receiver that will reject all stations except the one to which we wish to listen. As a minimum, this requires a tunable, highly selective bandpass filter. Unfortunately, such filters are not particularly easy to build, and since it is desirable to keep radio receivers as inexpensive as possible, an alternative approach is used.

Most AM radio receivers are of the superheterodyne type shown in Fig. 1.11. The main components of the superheterodyne receiver are (1) the radio frequency (RF) amplifier, which is tuned to the desired radio frequency; (2) the intermediate-frequency (IF) amplifier, which is fixed and provides most of the gain and selectivity; (3) the tunable local oscillator (LO); (4) the envelope detector; and (5) the audio-frequency (AF) amplifier, which matches the power level out of the envelope detector to that required by the speaker. The way this receiver works is that the desired radio signal is translated down to the IF center frequency by the tunable LO. The IF amplifier then rejects all other signals except the one radio station in its passband. Since the IF filter is fixed, the problem of building a tunable, highly selective filter is avoided. Further, since the carrier frequency supplied to the envelope detector is always the same, the envelope detector performance does not vary for different radio stations.

Commercial AM radio occupies the frequency band from 535 to 1605 kHz with radio stations located in the range 540 to 1600 kHz, spaced 10 kHz apart. Since double-sideband AM transmission is used, the baseband message bandwidth is 5 kHz. The bandwidths of the amplifiers in Fig. 1.11 are $10\text{kHz} < \text{BW}_{\text{RF}} < 910\text{kHz}$, $\text{BW}_{\text{IF}} = 10\text{kHz}$, and $\text{BW}_{\text{AF}} = 5\text{kHz}$, where BW_{RF} refers to the RF amplifier bandwidth; and so on. The choices of the IF and AF amplifier bandwidths are obvious, and the bandwidth of the RF amplifier will be justified shortly.

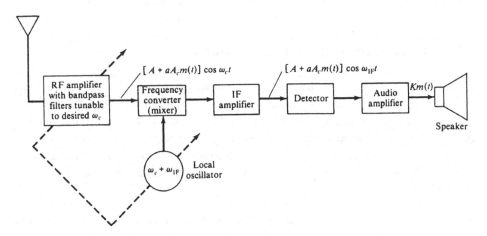

Fig. 1.11 Superheterodyne receiver

The center frequency of the IF amplifier is 455 kHz, and each carrier frequency from 540 to 1600 kHz must be shifted to this frequency. Thus the LO could be tuned over the range 995–2055 kHz or over the range 85–1145 kHz. The former tuning range was chosen for commercial application, since it constitutes only a 2:1 tuning ratio as opposed to the 13:1 ratio required by the latter range. As the LO is tuned, the center frequency of the RF amplifier is also varied. The purpose of the tunable RF amplifier is to reject the *image station.*

The concept of an image station is best illustrated by an example.

Example 1.6.1 We wish to select the radio station at 550 kHz. To do this, we tune the LO to $550 + 455 = 1005$ kHz. However, without an RF amplifier, the radio station at 1460 kHz will also be translated to the IF center frequency, since $1460 - 1005 = 455$. The station at 1460 kHz is called an *image station.* With the tunable RF amplifier, all stations outside the band $550 \pm \frac{1}{2}(\text{BW}_{\text{RF}})$ are eliminated before they get to the tunable LO. Thus the choice of the RF amplifier bandwidth is not critical as long as the image station is rejected, and the RF amplifier need not be highly selective, since all stations except the image station will be eliminated by the fixed IF filter.

Generalizing the discussion in the preceding example, we see that the center frequency of the RF amplifier is f_c, and the image station will be located at $f_c + 910$ kHz. Therefore, $\frac{1}{2}\text{BW}_{\text{RF}}$ only needs to be less than 910 kHz. Usually, the RF amplifier passband is stated to be $f_c + 455$ kHz, so $\frac{1}{2}\text{BW}_{\text{RF}} = 455$kHz. In practice, this value is not critical, since the mixer bandwidth is commonly less than this value. The minimum allowable bandwidth of the RF amplifier is determined by the bandwidth occupied by a single station, which is 10 kHz. We thus get the specification that 10 kHz $< \text{BW}_{\text{RF}} < 910$ kHz. Of course, since this filter must be tunable, it is not designed to have steep cutoff characteristics.

1.7   Quadrature Amplitude Modulation

In many applications, bandwidth is an important commodity. A widely used, bandwidth-efficient modulation technique called *quadrature amplitude modulation* (QAM) allows two message signals to be transmitted in the same frequency band without mutual interference. Given two distinct low-pass message signals $m_1(t)$ and $m_2(t)$ of bandwidth f_m hertz, message $m_1(t)$ amplitude modulates a carrier waveform $\cos \omega_c t$ and $m_2(t)$ modulates a carrier waveform $\sin\omega_c t$, and these two AMDSB-SC signals are summed to yield the QAM signal

$$s_{\text{QAM}}(t) = A_c m_1(t) \cos \omega_c t + A_c m_2(t) \sin \omega_c t. \qquad (1.7.1)$$

Fig. 1.12 QAM receiver block diagram

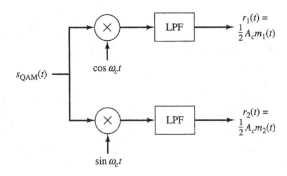

This waveform is called quadrature amplitude modulation, since the two message signals amplitude modulate separate carriers that are in phase quadrature with each other.

Clearly, $s_{QAM}(t)$ occupies no greater bandwidth than an AMDSB-SC signal for either message alone, so the information-carrying content of the available bandwidth has effectively been doubled.

A block diagram of a QAM receiver is shown in Fig. 1.12, where $s_{QAM}(t)$ is fed to two parallel branches. The upper branch multiplies by $\cos \omega_c t$ and the lower branch by $\sin \omega_c t$. After low-pass filtering these products in each branch, the output of the upper branch is $(A_c/2)m_1(t)$ and the output of the lower branch is $(A_c/2)m_2(t)$. Absolutely accurate frequency and phase tracking is essential to prevent distortion. For example, if the two local oscillators have phase errors of, say, θ_1 and θ_2 in the upper and lower branches, respectively, the demodulated outputs of the low-pass filters are

$$r_1(t) = \frac{A_c}{2}m_1(t) \cos \theta_1 + \frac{A_c}{2}m_2(t) \sin \theta_1 \qquad (1.7.2)$$

and

$$r_2(t) = \frac{A_c}{2}m_1(t) \sin \theta_2 + \frac{A_c}{2}m_2(t) \cos \theta_2. \qquad (1.7.3)$$

Thus phase errors in the local oscillators cause *crosstalk* between the two quadrature components.

Noncoherent demodulation of $s_{QAM}(t)$ is not possible since the envelope of $s_{QAM}(t)$ is $A_c\sqrt{m_1^2(t) + m_2^2(t)}$. Can some sort of transmitted carrier QAM be used, as in Eq. (1.4.12) for SSB, to facilitate envelope detection?

1.8   Frequency-Division Multiplexing

It often occurs that the particular transmission channel available to us, whether it is a wire pair, coaxial cable, or free space, will support a much wider bandwidth than that required by a single baseband message waveform. In these instances it is possible to employ *frequency-division multiplexing* (FDM) to allow simultaneous transmission of numerous unrelated message signals over the single communications channel. The FDM concept is illustrated in general by Fig. 1.13.

In a FDM system, each baseband message signal is transmitted using the same type of modulation (e.g., AMDSB-SC, conventional AM, SSB, or VSB), and the carrier frequencies, ω_1 through ω_N must be chosen such that none of the transmitted spectra overlap. Although the receiver could take a variety of forms, for the structure shown in Fig. 1.13, the bandpass filters are chosen to reject all messages except the one with the indicated carrier frequency. Demodulation then proceeds in a straightforward fashion.

The following example illustrates the details of FDM transmission.

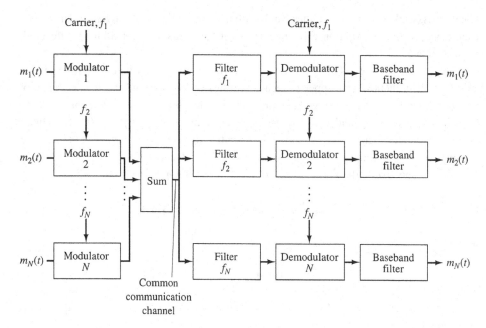

Fig. 1.13 Frequency-division multiplexing

Example 1.8.1 Four message signals, $m_1(t)$, $m_2(t)$, $m_3(t)$, and $m_4(t)$, with the Fourier transforms sketched in Fig. 1.14a are to be transmitted in the frequency range $2\pi(100,000) \le \omega \le 2\pi(120,000)$ over a single communications channel. To do this using FDM, we must select a modulation technique. Clearly, DSB transmission is not possible, since it would require a bandwidth of 32 kHz and only 20 kHz has been allocated. Although it is possible to use VSB here with properly chosen vestige bandwidths, we choose SSB modulation instead.

By using SSB transmission, the message signals will occupy only 16 kHz of the total 20 kHz available. We thus have some flexibility in choosing carrier frequencies. A simple choice would be to select carrier frequencies such that the transmitted messages are located contiguously at one end or the other of the allocated band. Instead, we will use the excess bandwidth to provide *guardbands*. Guardbands are vacant bands of frequencies between messages that "guard" against adjacent channel interference by providing extra spectral separation.

With this concept in mind and assuming USB SSB transmission, we assign the carrier frequencies $\omega_1 = 2\pi(100,500)$, $\omega_2 = 2\pi(105,500)$, $\omega_3 = 2\pi(110,500)$ and $\omega_4 = 2\pi(115,500)$. The multiplexed transmitted signal has the frequency content shown in Fig. 1.14b. We have thus allowed guardbands of 1000 Hz between adjacent messages, a

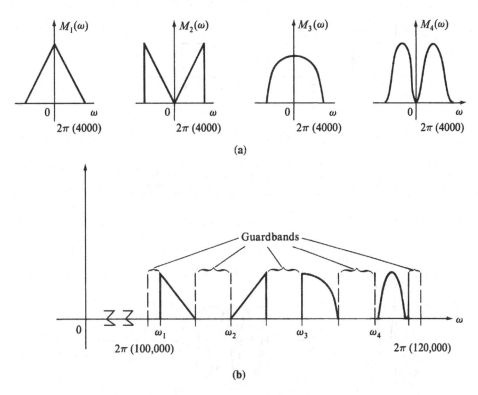

Fig. 1.14 Fourier transforms for Example 1.8.1: **a** Fourier transforms of baseband message signals; **b** frequency division multiplexing with guardbands

500-Hz guardband below message 1, and a 500-Hz guardband above message 4. Clearly, these choices are not unique and many other selections are possible.

1.9   Amplitude Shift Keying/On–Off Keying

Digital signals constitute an ever-growing portion of the message or information signals we are required to transmit over our communication systems. Many existing communication channels are inherently analog in nature, and some important communication channels, such as those found in the telephone network, have been designed primarily to carry analog signals. Of course, the telephone network and many other systems are evolving toward an all-digital or mostly digital configuration.

To transmit digital information over an analog channel, we use analog modulation methods. Although all types of AM modulation are employed in data transmission systems, we begin our discussion with AMDSB-SC. Suppose that we wish to send the binary sequence shown in Fig. 1.15a using the AMDSB-SC system represented by Fig. 1.15b.

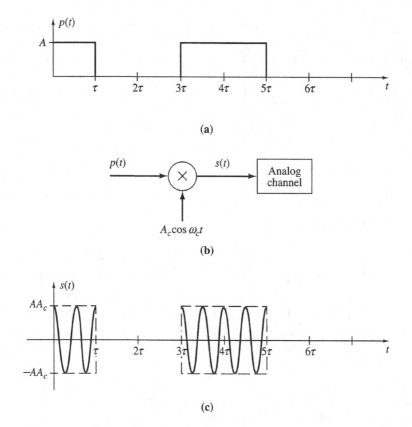

Fig. 1.15 Binary transmission using AMDSB-SC: **a** Binary message sequence; **b** AMDSB-SC system; **c** modulated carrier (OOK)

The input to the analog channel, $s(t)$, for this message signal is sketched in Fig. 1.15c. Thus, for each rectangular pulse in the original data stream, we get a pulsed version of the carrier with a rectangular envelope.

The frequency content of $s(t)$ is not as easy to ascertain. Certainly, if we have a single rectangular pulse multiplied by $A_c \cos \omega_c t$, we know that we have a sin x/x shape centered about $\pm \omega_c$. However, $p(t)$ is not just a single binary pulse, and there are some intervening spaces. Furthermore, we would like to analyze our system's frequency content for all possible data signals that might be sent, not just the single sequence currently being investigated. In general, whether we have a pulse or no pulse in any given interval is a random variable, and we must treat it as such in our analysis. We present the resulting spectral density in this section shortly.

To get some idea of the frequency content of a data signal transmitted using AMDSB-SC modulation, consider the periodic binary sequence sketched in Fig. 1.16a. The Fourier transform of this sequence is a series of spectral lines spaced $2\pi/2\tau = \pi/\tau$ rad/sec apart.

When $g(t)$ is passed through the AMDSB-SC system shown in Fig. 1.15b, the magnitude of the Fourier transform of $s(t)$ is as sketched in Fig. 1.16b. The envelope of the spectral lines depends on the pulse shape, while the spacing of the spectral lines depends on the period of the pulse train. When other pulse shapes are used, the time-domain signal in Fig. 1.15c and the frequency content in Fig. 1.16b both change accordingly.

The frequency content of a random sequence of pulses, with $q =$ probability of a pulse of amplitude A and $1 - q =$ probability of no pulse, can be shown to be here that $\mu_a = qA$ and $E\left[a_n^2\right] = qA^2$, so $\sigma_a^2 = qA^2(1 - q)$. Further,

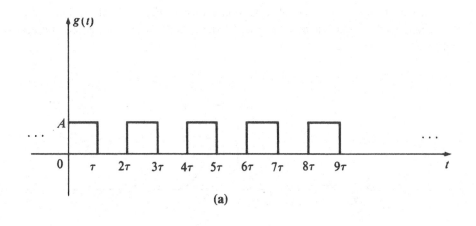

(a)

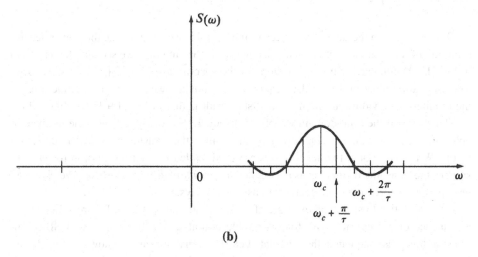

(b)

Fig. 1.16 Transmission of a periodic binary sequence using AMDSB-SC: **a** Periodic binary sequence; **b** frequency content of transmitted AMDSB-SC signal with message $g(t)$

$$T_s = \tau \quad \text{and} \quad P(\omega) = \tau e^{-j\omega\tau/2}\frac{\sin(\omega\tau/2)}{\omega\tau/2}.$$

so have the power spectral density of the pulse sequence,

$$S(\omega) = qA^2(1-q)\tau \left|\frac{\sin(\omega\tau/2)}{(\omega\tau/2)}\right|^2 + 2\pi q^2 A^2 \sum_{n=-\infty}^{\infty} \left|\frac{\sin n\pi}{n\pi}\right|^2 \delta\left(\omega - \frac{2n\pi}{\tau}\right)$$

$$= qA^2(1-q)\tau \left|\frac{\sin(\omega\tau/2)}{(\omega\tau/2)}\right|^2 + 2\pi q^2 A^2 \delta(\omega). \tag{1.9.1}$$

A sketch of this power spectral density is shown in Fig. 1.17. If the random pulse sequence is used to modulate a carrier, $\cos \omega_c t$, then copies of this power spectral density will be located about $\pm\omega_c$.

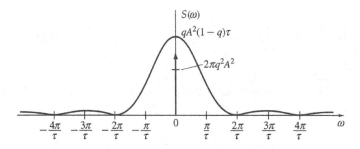

Fig. 1.17 Power spectral density of a random pulse sequence

To compare with the deterministic result in Fig. 1.16, we need to take the power density spectrum of the alternating pulse sequence in Fig. 1.16a, and then we see that the first term in Eq. (1.9.1) compares with the envelope of the spectral lines for $|\mathcal{F}\{g(t)\}|^2 = |G(\omega)|^2$. The alternating deterministic pulse sequence $g(t)$ thus allowed us to get a crude idea of the required bandwidth, but the more realistic result is that in Eq. (1.9.1) and Fig. 1.17.

The sketch of the time-domain waveform in Fig. 1.15c clearly indicates the motivation behind the nomenclature *on-off keying* (OOK), since the transmitted signal is either on or off. When more than two-level pulses are used, called *m*-ary transmission for *m*-level signals, the transmitted signal is usually called *amplitude shift keying* (ASK). The "keying" terminology is a remnant of telegraph transmission days.

It is instructive to sketch an example of ASK transmission of data. Figure 1.18a shows a sequence of pulses each with four allowable output levels. If we use AMDSB-SC to transmit this pulse sequence, the modulated carrier signal appears as shown in Fig. 1.18b. It is interesting to note that simply by observing *s(t)* in Fig. 1.18b alone, we cannot tell whether *s(t)* represents a four-level ASK sequence using AMDSB-SC modulation or a

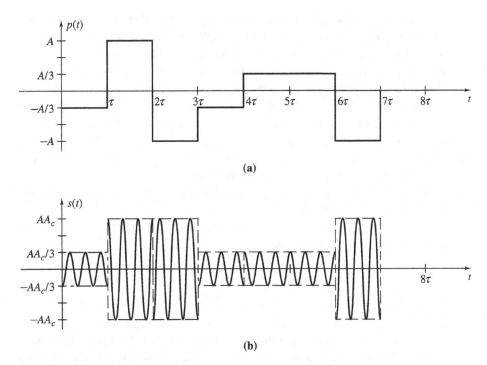

Fig. 1.18 Four-level transmission using AMDSB-SC: **a** Sequence of four-level pulses; **b** modulated carrier (ASK)

binary sequence using conventional AM modulation with a modulation index $a = 2/3$. Of course, when we design a communication system, the transmitter and receiver are jointly chosen and we do not have to guess what type of modulation was used.

The message sequence polarity in Fig. 1.18a is preserved in the phase of the carrier in Fig. 1.18b. Other AM methods are frequently used for data transmission.

Summary

In this chapter we have presented the fundamental principles of amplitude modulation, including AMDSB-SC, conventional AM, SSB, VSB, superheterodyne systems, and OOK/ASK. The effects of noise on AM modulation systems are discussed in Chap. 3.

Problems

1.1   A message signal, $m(t)$, is transmitted using AMDSB-SC modulation; thus the transmitted waveform is given by $s_{DSB}(t) = m(t) \cos \omega_c t$. During transmission, the

frequency and phase of the carrier signal are distorted, so that the received signal is
$r_{DSB}(t) = m(t) \cos[(\omega_c + \Delta\omega)t + \phi]$. If the LO signal is $A_c \cos \omega_c t$:

(a) Find an expression for the low-pass filter output of the receiver.

(b) If $\Delta\omega = 0$, find an expression for the total energy in the low-pass filter output,
and plot the energy as a function of ϕ for $\int_{-\infty}^{\infty} m^2(t)dt = 1$.

(c) Let $\phi = 0$, and describe the effect of the erroneous frequency reference. Sketch
a typical Fourier transform of the low-pass filter output. Assume a shape for
$M(\omega) = \mathcal{F}\{m(t)\}$.

1.2 Given the message signal $m(t) = 5 \sin 100\pi t$ and the carrier waveform
$2 \cos 2000\pi t$, sketch $s_{DSB}(t)$ and its Fourier transform.

1.3 For the message and carrier signals in Problem 1.2, sketch $s_{AM}(t)$ and its Fourier
transform if $a = 0.5$. See Eq. (1.3.1).

1.4 Show that for the conventional AM waveform in Eq. (1.3.1), the fraction of the total
average power in the sidebands is

$$\eta = \frac{a^2 A_c^2 \langle m^2(t) \rangle}{A^2 + a^2 A_c^2 \langle m^2(t) \rangle},$$

where $\langle m^2(t) \rangle$ is the average power in $m(t)$. η is called the *efficiency*.

1.5 Let $m(t) = \cos \omega_m t$ and plot η in Problem 1.4 as a function of the modulation index
a. Compare these results to the fraction of the total average power in the sidebands
for AMDSB-SC.

1.6 A conventional AM waveform can be generated by the switching modulator circuit
shown in Fig. 1.19. Assuming that $|m(t)|_{max} \ll A_c$, we have under

$$s_0(t) = \begin{cases} s_1(t), & \cos \omega_c t > 0 \\ 0, & \cos \omega_c t < 0. \end{cases}$$

Fig. 1.19 Switching
Modulator Circuit for Problem
1.6

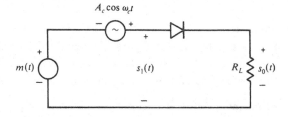

$$s_0(t) = \begin{cases} s_1(t), & \cos \omega_c t > 0 \\ 0, & \cos \omega_c t < 0. \end{cases}$$

Show how a conventional AM waveform can be obtained from $s_0(t)$ by appropriate filtering.

1.7 Consider the chopper modulator circuit shown in Fig. 1.20. An analysis of this circuit reveals that

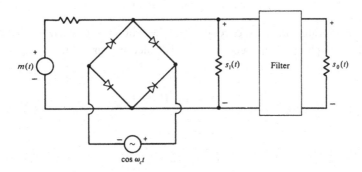

Fig. 1.20 Chopper Modulator Circuit for Problem 1.7

$$s_1(t) = \begin{cases} m(t), & \cos \omega_c t > 0 \\ 0, & \cos \omega_c t < 0. \end{cases}$$

Show that $s_0(t)$ can represent an AMDSB-SC signal. Specify the filter type, center frequency, cutoff frequencies, and bandwidth.

1.8 Consider the block diagram shown in Fig. 1.21. If $m(t)$ is the message signal and $y(t) = x^2(t)$, show how an AMDSB-SC wave can be obtained from $y(t)$. What happens if $y(t) = x^3(t)$?

Fig. 1.21 Modulator Block Diagram for Problem 1.8

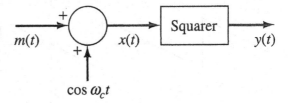

1.9 A nonlinear channel with input $x(t)$ and output $y(t)$ has the input/output relationship

$$y(t) = c_1 x(t) + c_2 x^2(t) + c_3 x^3(t).$$

If $x(t) = s_{DSB}(t) = m(t)A_c \cos \omega_c t$, can the AMDSB-SC wave be obtained undistorted at the receiver?

1.10 The signal

$$f(t) = \rho(t) \cos[\omega_c t + \theta(t)]$$

can be expanded using trigonometric identities to obtain

$$f(t) = \rho(t) \cos \theta(t) \cos \omega_c t - \rho(t) \sin \theta(t) \sin \omega_c t$$
$$= f_i(t) \cos \omega_c t - f_q(t) \sin \omega_c t.$$

$\rho(t)$ *and* $\theta(t)$ are called the envelope and phase of $f(t)$, respectively, while $f_i(t)$ is called the in-phase component and $f_q(t)$ is called the quadrature component. Show that

$$\rho(t) = \sqrt{f_i^2(t) + f_q^2(t)}$$

and

$$\theta(t) = \tan^{-1} \frac{f_q(t)}{f_i(t)}.$$

1.11 Sketch the output of an envelope detector for each of the input signals listed.
   (a) $f_1(t) = \sin t \cos 20{,}000t$.
   (b) $f_2(t) = [1 + \sin t] \cos 20{,}000t$.
   (c) $f_3(t) = \cos [20{,}000t - \cos t]$.
1.12 Given the periodic, normalized message waveform shown in Fig. 1.22:
   (d) Find the total average power in $m_n(t)$.
   (e) For a conventional AM waveform, what is the value of the modulation index that maximizes the efficiency η? (See Problem 1.4.)
   (f) Calculate the maximum value of η for the message signal $m_n(t)$.

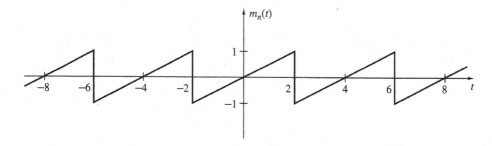

Fig. 1.22 Periodic Waveform for Problem 1.12

1.13 Consider the message, modulator, and channel shown in Fig. 1.23a. Sketch the Fourier transform at each labeled point in the demodulators in Fig. 1.23b, c.

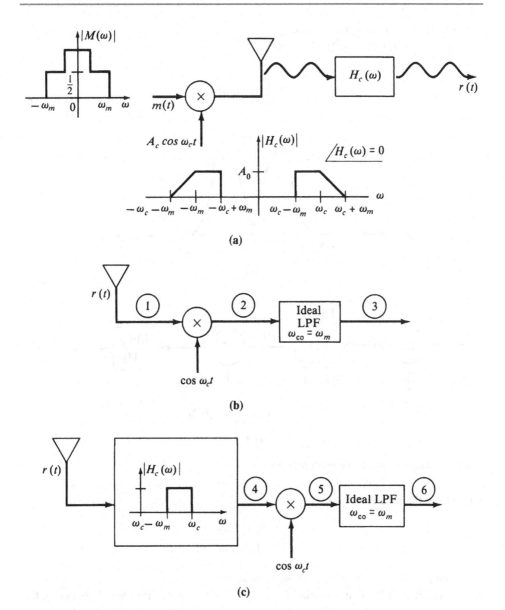

Fig. 1.23 Modulator and Demodulators for Problem 1.13

1.14 For $m(t)$ with the Fourier transform shown in Fig. 1.24a, show that the block diagram in Fig. 1.24b functions as a data scrambler.

1.15 Let $g_h(t)$ be the Hilbert transform of the function $g(t)$. Prove the following properties.

   (a) If $g(t)$ is even, $g_h(t)$ is odd.

(b) If $g(t)$ is odd, $g_h(t)$ is even.

(c) $\int_{-\infty}^{\infty} g^2(t)\, dt = \int_{-\infty}^{\infty} g_h^2(t)\, dt$.

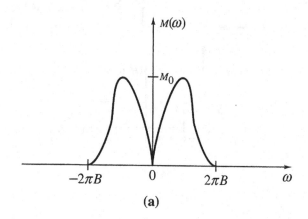

(a)

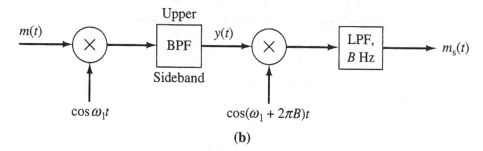

(b)

Fig. 1.24 Data Scrambler Diagram for Problem 1.14

1.16 Given the time function

$$m(t) = \begin{cases} 1, & |t| \le 1 \\ 0, & \text{otherwise,} \end{cases}$$

   find $m_h(t) = \mathcal{H}\{m(t)\}$. Sketch $m(t)$ and $m_h(t)$.

1.17 Given that SSB-TC modulation is used to transmit $m(t)$ in Problem 1.16, sketch the output of an envelope detector. Contrast this result with coherent demodulation of a SSB-SC wave carrying the same $m(t)$.

1.18 State the principal advantage of each of the following modulation methods.

   (d) Conventional AM.

   (e) AMDSB-SC.

(f) AMSSB-SC.

1.19 (This problem was provided by J. L. LoCicero.) Given the block diagram of a SSB phase shift modulator, sometimes called a *Hartley modulator*, in Fig. 1.25, show that this system can generate an upper- or lower-sideband SSB wave for a general message signal $m(t)$.

Hint: Use Eqs. (1.4.3)–(1.4.5).

1.20 (This problem was provided by J. L. LoCicero.) For a general $m(t)$, show that the Weaver modulator in Fig. 1.26 can be used to generate an upper- or lower-sideband SSB wave.

Hint: Use Eqs. (1.4.3)–(1.4.5).

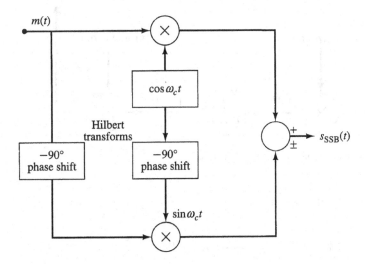

Fig. 1.25 Hartley Modulator for Problem 1.19

1.21 Noting that Eq. (1.4.3) can be written as $M_h(\omega) = -jM(\omega)\,\text{sgn}\,(\omega)$, derive Eq. (1.4.2).

1.22 A periodic signal is given by

$$f(t) = \sum_{n=-\infty}^{\infty} (-1)^n p(t-nT),$$

where $p(t)$ is a rectangular pulse of amplitude A and length T seconds.

(a) Write an expression for the Hilbert transform of $f(t)$, denoted $f_h(t)$.

(b) Sketch $f(t)$ and $f_h(t)$.

(c) If $f(t)$ is transmitted by SSB-SC and the demodulator local oscillator has a phase error of $\theta°$, what is the demodulator output?

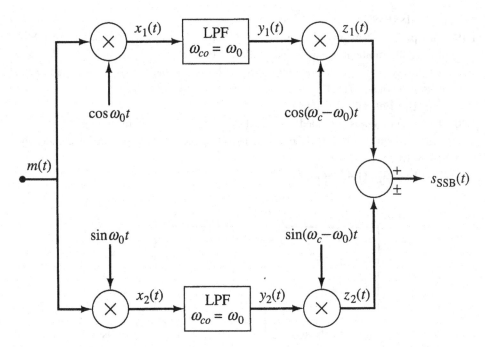

Fig. 1.26 Weaver Modulator for Problem 1.20

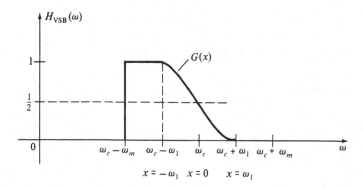

Fig. 1.27 VSB Filter for Problem 1.23

1.23 The complementary symmetry requirement of VSB can be specified in a variety of ways. Referring to Fig. 1.27, one way we can specify this complementary symmetry is by considering only positive ω, and defining a new variable, say x, with its origin at $\omega = \omega_c$. Letting $G(x)$ be the filter shape about $x = 0$, we have for the complementary symmetry requirement that $|G(-\alpha)| = 1 - |G(\alpha)|$. Verify that the function

$$G(x) = \begin{cases} \frac{1}{2}\left[1 - \sin\frac{\pi x}{2\omega_1}\right], & |x| \leq \omega_1 \\ 0, & x > \omega_1 \\ 1, & -\omega_m \leq x < -\omega_1, \end{cases}$$

satisfies the VSB complementary symmetry requirements.

1.24 It is also possible to represent the VSB complementary symmetry requirement as a sum of an ideal filter response and a filter with odd symmetry about ω_c, as shown in Fig. 1.28. Show that we can write a time-domain expression for a VSB wave of the form

$$s_{VSB}(t) = m(t)\cos\omega_c t \pm \left[m_h(t) + m_\beta(t)\right]\sin\omega_c t.$$

Specify $m_\beta(t)$.

1.25 Using the expression for $s_{VSB}(t)$ given in Problem 1.24, demonstrate that coherent demodulation can regain $m(t)$. Exhibit what can happen if the local oscillator has an inaccurate frequency or phase reference.

1.26 Starting with the expression for $s_{VSB}(t)$ in Problem 1.24, show that envelope detection of VSB can be effective if a suitable carrier term is injected or transmitted. State the required assumptions.

1.27 For broadcast AM radio, show that image stations can be eliminated without RF filtering by an appropriate choice of the IF center frequency.

1.28 We wish to receive the AM radio station at 980 kHz. What is the local oscillator frequency? What is the frequency of the image station? What is the center frequency of the tunable RF amplifier?

1.29 The sequence of pulses shown in Fig. 1.29 is to be transmitted using AMDSB-SC. If the equation for the pulse shape when centered at $t = 0$ is $p(t) = V\cos(\pi t/\tau)$ for $|t| \leq \tau/2$, and 0 otherwise, sketch the modulated carrier. Assume that $\omega_c \gg 2\pi/\tau$.

1.30 If the message to be transmitted using AMDSB-SC is

$$f(t) = \sum_{n=-\infty}^{\infty} p(t - 2n\tau),$$

where $p(t)$ is given in Problem 1.29, sketch the frequency content of the transmitted waveform. Assume that $\omega_c \gg 2\pi/\tau$.

1.31 A three-level pulse sequence shown in Fig. 1.30 is to be transmitted using AMDSB-SC. Sketch the transmitted waveform. Assume that $\omega_c \gg 2\pi/\tau$.

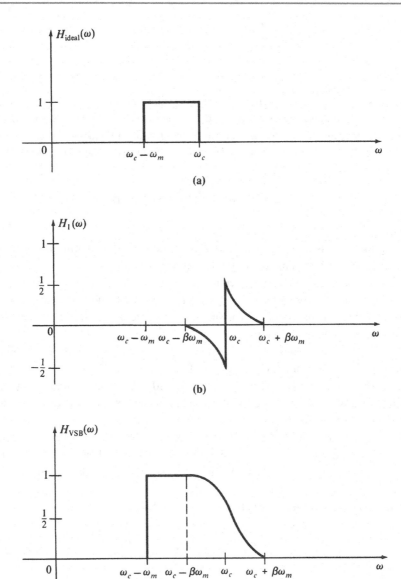

Fig. 1.28 VSB Filter Construction for Problem 1.24

1.32 Five message signals, $m_i(t)$, $i = 1, 2, 3, 4, 5$, each with a low-pass bandwidth of 5 kHz, are to be transmitted using frequency-division multiplexing in the band $2\pi(100,000) \leq \omega \leq 2\pi(130,000)$. Can AMDSB methods be used? Why or why not? If lower-sideband AMVSB transmission is used, specify the carrier frequency

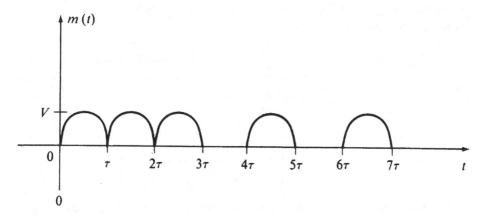

Fig. 1.29 Pulse Sequence for Problem 1.29

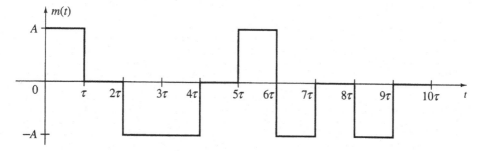

Fig. 1.30 Three Level Pulse Sequence for Problem 1.31

and the band of frequencies occupied for each message assuming that no bandwidth is allocated to guardbands.

1.33 Spread spectrum communication systems reduce the effects of narrow bandwidth interference by spreading the bandwidth of the transmitted signal out over a range of frequencies much greater than the bandwidth occupied by an AMDSB-SC signal alone. Consider the AMDSB-SC signal in Eq. (1.2.1), where $m(t)$ is the message with $|M(\omega)|$ as in Fig. 1.1a.

Let $p(t)$ be a sequence of ± 1 pulses with $\mathcal{F}\{p(t)\} = P(\omega)$ as in Fig. 1.31, where $\omega_p \gg \omega_c \gg \omega_m$. The spreading of the bandwidth is accomplished by multiplying by $p(t)$ to obtain $s_{DSB}(t)p(t)$.

(a) Find the band of frequencies occupied by $s_{DSB}(t)p(t)$.

(b) If there is no distortion, the transmitted signal $s_{DSB}(t)p(t)$ appears at the receiver input. Show that multiplying by $p(t)$ again produces $S_{DSB}(t)$, from which we can obtain the message $m(t)$.

(c) Assume that there is additive distortion $n(t)$ with $\mathcal{F}\{n(t)\} = N(\omega)$ such that

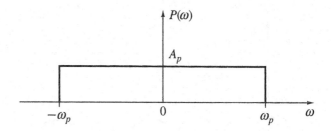

Fig. 1.31 Pulse Fourier
Transform for Problem 1.33

$$N(\omega) = \begin{cases} N_0, & \omega_c - \omega_m \leq |\omega| < \omega_c + \omega_m \\ 0, & \text{elsewhere.} \end{cases}$$

Show that if the received waveform is $r(t) = s_{\text{DSB}}(t)p(t) + n(t)$, then multiplying by $p(t)$ at the receiver reduces the noise power in the bandwidth of interest by a factor proportional to $1/\omega_p$.

Phase and Frequency Modulation

2

2.1 Introduction

Amplitude modulation is often called *linear modulation,* since superposition applies for AMDSB-SC systems. For instance, in AMDSB-SC if a message signal $m_1(t)$ generates a modulated waveform $s_2(t)$, and message $m_2(t)$ generates the waveform $s_2(t)$, the sum of the messages, $m_1(t) + m_2(t)$, will produce the modulated waveform $s_1(t) + s_2(t)$. If we ignore the constant added to the message signal, AMDSB-TC, or conventional AM, also has this property. In this chapter we turn our attention to angle modulation, where instead of varying the carrier amplitude in proportion to the message, we vary the carrier phase or frequency in relation to the message signal. Since the phase and frequency are in the argument of a sine or cosine function, angle modulation does not obey superposition, and hence it is sometimes called *nonlinear modulation.* This nonlinearity will become clear as the development progresses.

2.2 Phase and Frequency Modulation

In *phase modulation* (PM) we vary the phase of the carrier signal in a linear relationship with the message signal, $m(t)$. That is, given the carrier signal

$$s_{\text{PM}}(t) = A_c \cos[\omega_c t + \theta(t)], \tag{2.2.1}$$

we choose

$$\theta(t) = c_p m(t), \tag{2.2.2}$$

© The Author(s), under exclusive license to Springer Nature Switzerland AG 2023
J. D. Gibson, *Analog Communications*, Synthesis Lectures on Communications,
https://doi.org/10.1007/978-3-031-19584-6_2

where c_p is a constant. Certainly, Eqs. (2.2.1) and (2.2.2) do not represent a linear modulation method, since we can rewrite Eq. (2.2.1) as

$$s_{PM}(t) = A_c \cos\big[c_p m(t)\big] \cos \omega_c t - A_c \sin\big[c_p m(t)\big] \sin \omega_c t \qquad (2.2.3)$$

by using a trigonometric identity. The carrier frequency terms are thus modulated by nonlinear transformations of $c_p m(t)$. Equation (2.2.3) is not the best way to visualize what the time-domain waveform of a phase-modulated signal looks like, however. In fact, in order to sketch the time-domain waveform in all but the simplest cases, we need to introduce the concept of *frequency modulation* (FM).

When first considering frequency modulation, we are led immediately to the problem of what we mean by the frequency of a waveform. Intuitively, the frequency of a sine or cosine waveform makes sense only if the frequency is fixed (constant) and the sine or cosine wave exists for all time. This problem arises primarily because it only makes sense to talk about the sine or cosine of an angle (not a frequency), and generally we consider the angle to vary linearly with time. For example, in

$$s(t) = A_c \cos\phi(t) = A_c \cos[\omega_c t + \psi], \qquad (2.2.4)$$

where ψ is a constant, the angle $\phi(t)$ varies in linear proportion to the fixed radian frequency ω_c. Thus, in this specific case, we can interpret the radian frequency as the derivative of the angle; that is, $\omega_c = d\phi(t)/dt$, and we get an answer that coincides with our intuition.

If $\phi(t)$ does not vary linearly with time, the justification for defining the frequency as the derivative of the angle is not so clear. Since we are going to be talking about signals whose frequency is changing with time, we must have a definition of frequency that allows us to speak of the frequency of a signal at any time instant. For this reason we define the *instantaneous radian frequency,* denoted by ω_i, as

$$\omega_i \triangleq \frac{d}{dt}\phi(t), \qquad (2.2.5)$$

where $\phi(t)$ is the angle of the carrier waveform. If $\phi(t) = \omega_c t + \psi$, this definition agrees with our intuition. Otherwise, however, the result may not be so intuitive. For Eqs. (2.2.1) and (2.2.2), we have

$$\omega_i = \frac{d}{dt}\big[\omega_c t + c_p m(t)\big] = \omega_c + c_p \frac{d}{dt} m(t). \qquad (2.2.6)$$

This result implies that for PM, the instantaneous frequency varies about the fixed value ω_c in linear proportion to the derivative of the message signal.

A logical definition of a frequency-modulated waveform would be to have the instantaneous frequency vary linearly with respect to $m(t)$ itself; hence

$$\omega_i = \omega_c + c_f m(t), \tag{2.2.7}$$

where c_f is a constant, represents the instantaneous frequency of an FM signal. To write a time-domain expression for the FM waveform, we note Eq. (2.2.5) and integrate both sides of Eq. (2.2.7) to find

$$\phi(t) = \int \omega_i \, dt = \omega_c t + c_f \int m(t) \, dt + \psi, \tag{2.2.8}$$

where ψ represents a constant of integration. Note that it is customary to write the instantaneous frequency as ω_i rather than $\omega_i(t)$, even though generally it is a function of time. Using Eq. (2.2.8), we can write the FM signal as

$$s_{FM}(t) = A_c \cos\left[\omega_c t + c_f \int m(t) \, dt\right], \tag{2.2.9}$$

where ψ is chosen to be zero for simplicity.

By comparing Eqs. (2.2.1) and (2.2.2) with Eq. (2.2.9), it is evident that FM and PM are different modulation methods, but it is equally evident that they are both angle modulation methods and hence closely related. The following examples illustrate the instantaneous frequency concept and the relationship between FM and PM.

Example 2.2.1 Consider the message signal sketched in Fig. 2.1a. This message can be transmitted by both FM and PM. For PM let $c_p = \pi/4 \, \text{rad/V}$ and assume for this example that $T = 2T_c$, where $\omega_c = 2\pi/T_c$. Under these assumptions, the PM waveform can be sketched as shown in Fig. 2.1b.

To transmit m(t) using FM, let $c_f = \pi/2T$ rad/s per volt, so that

$$\omega_i = \omega_c + \frac{\pi}{2T} m(t) \tag{2.2.10}$$

or

$$\phi(t) = \omega_c t + \frac{\pi}{2T} \int m(t) \, dt. \tag{2.2.11}$$

We can sketch $s_{FM}(t)$ by referring to either Eq. (2.2.10) or (2.2.11). From Eq. (2.2.10) we see that the instantaneous frequency is changing between ω_c and $\omega_c + \pi/2T = 2\pi/T_c + \pi/4T_c = \frac{9}{8}\omega_c$. Thus $s_{FM}(t)$ can be sketched as shown in Fig. 2.1c. Note that the same result is evident from Eq. (2.2.11), which implies that the phase linearly increases when $m(t)$ is nonzero up to 90° in the interval $T \le t \le 2T$, and in every interval where $m(t)$ is 1.

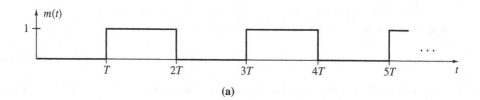

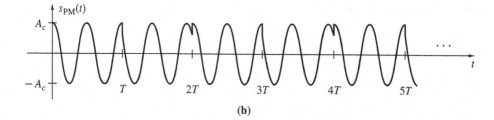

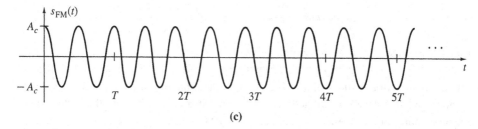

Fig. 2.1 Waveforms for Example 2.2.1: **a** Message signal for Example 2.2.1; **b** PM waveform for $m(t)$ in part **a**; **c** FM waveform for $m(t)$ in part **a**

Example 2.2.2 Given the waveform

$$s(t) = 10 \cos\left[90 \times 10^6 t + 200 \cos 2000t\right],$$

(a) If $s(t)$ is a PM signal, what is $m(t)$? From Eqs. (2.2.1) and (2.2.2) we see immediately that $c_p m(t) = 200 \cos 2000t$, so that

$$m(t) = \left(\frac{200}{c_p}\right) \cos 2000t.$$

(b) If $s(t)$ is an FM signal, what is $m(t)$? Following Eq. (2.2.5), we obtain

$$\omega_i = 90 \times 10^6 - 4 \times 10^5 \sin 2000t,$$

which by comparison with Eq. (2.2.7) reveals that $c_f m(t) = -4 \times 10^5 \sin 2000t$ or $m(t) = \left(-4 \times 10^5 / c_f\right) \sin 2000t$.

Using the phase and instantaneous frequency, we can thus determine and sketch (if we so desire) the corresponding FM and PM time-domain waveforms for a given message, but we have not yet specified the frequency-domain representation of an angle-modulated signal. This is the topic of the following section.

2.3 Bandwidth Requirements

To begin the investigation of FM bandwidth requirements, we restrict the message signal to be a pure tone given by

$$m(t) = A \cos \omega_m t. \qquad (2.3.1)$$

Although virtually all messages of interest are much more complicated than $m(t)$ in Eq. (2.3.1), we shall discover that the bandwidth requirements of an FM wave carrying this $m(t)$ are representative of the bandwidth requirements of many other physically important signals. By assuming a message signal of this form, we can thus illustrate several critical concepts without undue mathematical difficulty.

If we use $m(t)$ in Eq. (2.3.1) to frequency modulate a carrier signal, the instantaneous frequency of the carrier can be expressed as

$$\omega_i = \omega_c + \Delta\omega \cos \omega_m t, \qquad (2.3.2)$$

where $\Delta\omega$ is a constant such that $\Delta\omega \ll \omega_c$. The instantaneous frequency of the FM wave thus varies about the fixed carrier frequency ω_c at a rate of ω_m rad/s with a maximum frequency change of $\Delta\omega$ rad/s. The quantity $\Delta\omega$ is called the *maximum frequency deviation* or just the *frequency deviation* of the FM waveform. Using Eq. (2.2.8), we find that

$$\phi(t) = \int \omega_i \, dt = \omega_c t + \frac{\Delta\omega}{\omega_m} \sin \omega_m t + \psi, \qquad (2.3.3)$$

where we normally let $\psi = 0$. A time-domain equation for the FM signal is therefore

$$s_{FM}(t) = A_c \cos\left[\omega_c t + \frac{\Delta\omega}{\omega_m} \sin \omega_m t\right]$$
$$= A_c \cos[\omega_c t + \beta \sin \omega_m t], \qquad (2.3.4)$$

where we have defined

$$\beta = \frac{\Delta\omega}{\omega_m}, \qquad (2.3.5)$$

which is called the *modulation index*. Since β depends on $\Delta\omega$ and ω_m, and $\Delta\omega$ indicates the frequency range over which the FM signal will vary and ω_m indicates how rapidly this variation takes place, β is clearly related to the required bandwidth of the FM wave.

Two different classes of FM waves are of interest, depending on the value of β. For $\beta \ll \pi/2$ (usually, $\beta < 0.2$ or 0.5), called *narrowband FM*, we use trigonometric identities to write

$$
\begin{aligned}
s_{FM}(t) = A_c \cos\omega_c t \cos[\beta \sin\omega_m t] \\
- A_c \sin\omega_c t \sin[\beta \sin\omega_m t],
\end{aligned}
\tag{2.3.6}
$$

which for β small we can approximate as

$$
s_{FM}(t) \cong A_c \cos\omega_c t - A_c\beta \sin\omega_m t \sin\omega_c t,
\tag{2.3.7}
$$

since $\cos[\beta \sin\omega_m t] \cong 1$ and $\sin[\beta \sin\omega_m t] \cong \beta \sin\omega_m t$. Using a trigonometric identity on the second term in Eq. (2.3.7) reveals that the bandwidth of a narrowband FM wave is approximately $2\omega_m$, the same as AMDSB methods.

When $\beta > \pi/2$ the frequency-modulated signal is called *wideband FM*, which is the type of FM used most often in analog communication systems. To determine the bandwidth requirements of wideband FM, we build on Eq. (2.3.6) by expanding the terms $\cos[\beta \sin\omega_m t]$ and $\sin[\beta \sin\omega_m t]$ in a Fourier series. Since $\text{Re}\{e^{j\beta \sin\omega_m t}\} = \cos[\beta \sin\omega_m t]$ and $\text{Im}\{e^{j\beta \sin\omega_m t}\} = \sin[\beta \sin\omega_m t]$, we can find both expressions by writing a Fourier series for

$$
g(t) = e^{j\beta \sin\omega_m t}, \quad -\frac{T}{2} \le t \le \frac{T}{2},
\tag{2.3.8}
$$

where $T = 2\pi/\omega_m$.

Straightforwardly, the complex Fourier series coefficients are given by

$$
c_n = \frac{1}{T} \int_{-T/2}^{T/2} e^{j(\beta \sin\omega_m t - n\omega_m t)} dt = \frac{1}{2\pi} \int_{-\pi}^{\pi} e^{j(\beta \sin x - nx)} dx
\tag{2.3.9}
$$

by making the change of variable $x = \omega_m t$. The integral in Eq. (2.3.9) can be evaluated only by means of an infinite series, but fortunately, because it often appears in many physical problems, it is tabulated extensively. The integral in Eq. (2.3.9) is called the *Bessel function of the first kind* and is represented by

$$
J_n(\beta) = \frac{1}{2\pi} \int_{-\pi}^{\pi} e^{j(\beta \sin x - nx)} dx,
\tag{2.3.10}
$$

which is a function of both n and β. The functions in Eq. (2.3.10) have the properties that

$$
J_n(\beta) = J_{-n}(\beta) \quad \text{for } n \text{ even}
\tag{2.3.11a}
$$

and

$$J_n(\beta) = -J_{-n}(\beta) \quad \text{for } n \text{ odd.} \tag{2.3.11b}$$

Plots of the Bessel functions of the first kind are shown in Fig. 2.2.

Writing out the Fourier series term by term and employing Eqs. (2.3.11a, b) to combine the positive and negative terms with equal magnitudes of n yields

$$e^{j\beta \sin \omega_m t} = J_0(\beta) + 2[J_2(\beta) \cos 2\omega_m t + J_4(\beta) \cos 4\omega_m t + \cdots]$$
$$+ j2[J_1(\beta) \sin \omega_m t + J_3(\beta) \sin 3\omega_m t + \cdots]. \tag{2.3.12}$$

Using Euler's identity on the left side of Eq. (2.3.12) and equating real and imaginary parts produces the desired expressions,

$$\cos[\beta \sin \omega_m t] = J_0(\beta) + 2J_2(\beta) \cos 2\omega_m t + 2J_4(\beta) \cos 4\omega_m t + \cdots$$
$$= J_0(\beta) + 2 \sum_{\substack{n=2 \\ n \text{ even}}}^{\infty} J_n(\beta) \cos n\omega_m t \tag{2.3.13}$$

and

$$\sin[\beta \sin \omega_m t] = 2J_1(\beta) \sin \omega_m t + 2J_3(\beta) \sin 3\omega_m t + \cdots$$

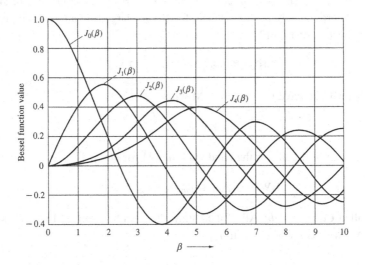

Fig. 2.2 Plots of Bessel functions

$$= 2 \sum_{\substack{n=1 \\ n \text{ odd}}}^{\infty} J_n(\beta) \sin n\omega_m t. \tag{2.3.14}$$

Substituting Eqs. (2.3.13) and (2.3.14) into Eq. (2.3.6) and using trigonometric identities gives

$$s_{FM}(t) = A_c\{J_0(\beta) \cos \omega_c t - J_1(\beta)[\cos(\omega_c - \omega_m)t - \cos(\omega_c + \omega_m)t]$$
$$+ J_2(\beta)[\cos(\omega_c - 2\omega_m)t + \cos(\omega_c + 2\omega_m)t]$$
$$- J_3(\beta)[\cos(\omega_c - 3\omega_m)t - \cos(\omega_c + 3\omega_m)t] + \cdots\}$$
$$= A_c\left\{ J_0(\beta) \cos \omega_c t + \sum_{n=1}^{\infty} (-1)^n J_n(\beta)[\cos(\omega_c - n\omega_m)t \right.$$
$$\left. + (-1)^n \cos(\omega_c + n\omega_m)t]\right\}. \tag{2.3.15}$$

Since for a specific application β is fixed, it is evident from Eq. (2.3.15) that the FM wave consists of a carrier plus an infinite number of sidebands spaced at radian frequencies $\pm\omega_m, \pm2\omega_m, \ldots$, about the carrier. In comparison to AMDSB methods with the same message signal, which requires a bandwidth of only $2\omega_m$, we see that the nonlinear angle modulation has generated numerous additional sidebands.

Equation (2.3.15) is valid for all β, and although there are in general an infinite number of sidebands, the number of *significant* sidebands depends on the value of β. More specifically, for $n > 1$ and $\beta \ll n$, $J_n(\beta) \cong 0$, but for $\beta \gg 1$, the number of significant sidebands is approximately $n = \beta$. Hence, for β very large, the approximate bandwidth requirement for (wideband) FM is

$$BW \cong 2n\omega_m = 2\beta\omega_m = 2\Delta\omega, \tag{2.3.16}$$

while for β small we have narrowband FM with

$$BW \cong 2\omega_m. \tag{2.3.17}$$

A general rule of thumb for the bandwidth of FM for any β is

$$BW = 2(\Delta\omega + \omega_m) = 2\omega_m(1 + \beta), \tag{2.3.18}$$

which is called *Carson's rule.* All of the bandwidths can be expressed in hertz simply by replacing $\Delta\omega$ with Δf and ω_m with f_m.

Example 2.3.1 In broadcast FM, the FCC restricts Δf to be 75 kHz and $(f_m)_{\max} = $ maximum frequency in $m(t)$ to be less than or equal to 15 kHz. The smallest value of the modulation index for $\Delta f = 75$ kHz is

$$\beta = \frac{\Delta f}{(f_m)_{\max}} = \frac{75\text{kHz}}{15\text{kHz}} = 5. \tag{2.3.19}$$

For message signals with lower values of f_m, β will be larger. Using $\beta = 5$ and Carson's rule, we find that

$$\text{BW} = 2f_m(1 + \beta) = 180\,\text{kHz}. \tag{2.3.20}$$

Since the total bandwidth allocated to each FM station is 200 kHz, we are safely within the specified band. Note that for smaller values of f_m, we shall have a large β, or wideband FM, so then $\text{BW} \cong 2\,\Delta f = 150\,\text{kHz}$.

Example 2.3.2 Given the angle-modulated waveform

$$s(t) = 10\cos\left[90 \times 10^6 t + 200\cos 2000t\right],$$

what is its required bandwidth? Whether $s(t)$ represents a PM or FM wave, the bandwidth can be found from Carson's rule. The instantaneous frequency is given by

$$\omega_i = 90 \times 10^6 - 4 \times 10^5 \sin 2000t,$$

so $\omega_m = 2000$ and $\Delta\omega = 4 \times 10^5$. Therefore,

$$\text{BW} = 2\left[4 \times 10^5 + 2000\right] = 8.04 \times 10^5 \text{rad/s}.$$

Note that $\beta = 4 \times 10^5 / 2 \times 10^3 = 200$, which is wideband FM, hence

$$\text{BW} \cong 2\Delta\omega = 8 \times 10^5 \text{rad/s}.$$

Although all of the results in this section concerning the bandwidth of an FM signal have been derived under the assumption of a single-tone modulating signal, we are fortunate in that the expressions can be extended to yield bandwidth guidelines for more general messages. In particular, if we wish to use Eqs. (2.3.16)–(2.3.18) to determine the bandwidth of a general message signal $m(t)$ with a highest frequency present in $m(t)$ of $2\pi W_m$, then we need only let $\omega_m = 2\pi W_m$ and $\Delta\omega = c_f[\max |m(t)|]$, for which we find

$$\beta = \frac{\Delta\omega}{\omega_m} = \frac{c_f[\max |m(t)|]}{2\pi W_m}. \tag{2.3.21}$$

Of course, Eqs. (2.3.16)–(2.3.18) are only approximations to the bandwidth of any FM signal, and other rules of thumb are found in the literature.

2.4   Modulation and Demodulation Methods

Just as in developing bandwidth requirements, modulation methods for FM and PM can be separated into narrowband and wideband categories. Since we considered only single-tone messages in Sect. 2.3, we need to illustrate the narrowband concepts for general messages and both PM and FM. To begin, we can write a general angle-modulated signal as

$$s(t) = A_c \cos[\omega_c t + \phi(t)], \tag{2.4.1}$$

which can then be expanded using a trigonometric identity as

$$s(t) = A_c \cos \phi(t) \cos \omega_c t - A_c \sin \phi(t) \sin \omega_c t. \tag{2.4.2}$$

If $|\phi(t)|_{\max}$ is small, then Eq. (2.4.2) is approximately

$$s(t) = A_c \cos \omega_c t - A_c \phi(t) \sin \omega_c t. \tag{2.4.3}$$

For PM, we have that $\phi(t) = c_p m(t)$, so the condition for narrowband PM is $c_p |m(t)|_{\max} \ll \pi/2$, and the narrowband PM wave is

$$s_{\text{PM}}(t) = A_c \cos \omega_c t - A_c c_p m(t) \sin \omega_c t. \tag{2.4.4}$$

For FM, $\phi(t) = c_f \int m(t) dt$, and the FM waveform is

$$s_{\text{FM}}(t) = A_c \cos \omega_c t - A_c c_f \left[\int m(t) dt \right] \sin \omega_c t, \tag{2.4.5}$$

where $c_f |\int m(t) dt|_{\max} \ll \pi/2$. From Eqs. (2.4.4) to (2.4.5), we can implement narrowband PM or FM as illustrated by the block diagram in Fig. 2.3.

The generation of wideband FM can be separated into two types: indirect FM and direct FM. *Indirect FM* starts with a narrowband frequency-modulated signal and uses frequency multiplication to generate a wideband FM signal. This procedure is illustrated by Fig. 2.4. If the narrowband FM signal is

$$x(t) = A_c \cos[\omega_c t + \phi(t)],$$

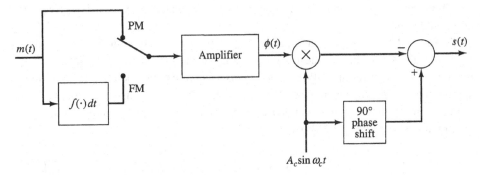

Fig. 2.3   Generation of narrowband PM or FM

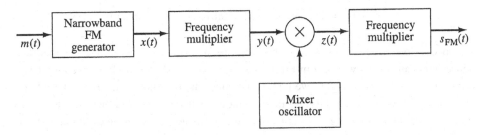

Fig. 2.4   Indirect generation of wideband FM

the signal after frequency multiplication is

$$y(t) = A_c \cos[n\omega_c t + n\phi(t)],$$

so that the instantaneous frequency of $y(t)$ is

$$\omega_i = n\omega_c + n\frac{d}{dt}\phi(t) = n\omega_c + nc_f m(t).$$

As a result, the new frequency deviation is n times the narrowband signal's frequency deviation, and thus a wideband FM wave is generated. The frequency multiplication also increases the carrier frequency by a factor of n. If this is not desirable, the carrier frequency of the frequency multiplier output can be adjusted by the mixer oscillator as shown in Fig. 2.4. Of course, the mixing operation only relocates the wideband FM waveform in the frequency spectrum; it does not affect the frequency deviation.

For *direct FM* methods, the modulating signal directly varies the carrier frequency. Direct FM may also require some frequency multiplication, but it is usually less than in indirect FM. There are a variety of methods for directly controlling the carrier frequency with the modulating signal, including varying the inductance or capacitance of a resonant circuit, using a voltage-controlled oscillator (VCO), or changing the switching frequency of a multivibrator.

To demodulate FM signals, it is necessary to build a device or system which produces an output amplitude that is linearly proportional to the instantaneous frequency of the FM wave. Such devices are called *frequency discriminators,* and there are many different approaches to their implementation. An ideal differentiator has a linear amplitude versus frequency magnitude response and hence should be suitable for FM demodulation.

To see this, consider the FM wave

$$s_{\text{FM}}(t) = A_c \cos\left[\omega_c t + c_f \int m(t)dt\right], \tag{2.4.6}$$

which when passed through an ideal differentiator becomes

$$\frac{d}{dt}s_{\text{FM}}(t) = -A_c\left[\omega_c + c_f m(t)\right]\sin\left[\omega_c t + c_f \int m(t)dt\right]. \tag{2.4.7}$$

Since $c_f|m(t)|_{\max} = \Delta\omega \ll \omega_c$, the envelope of Eq. (2.4.7) is always positive, and hence $m(t)$ can be recovered from Eq. (2.4.7) by envelope detection. For envelope detection to be effective, it is imperative that A_c be constant. To achieve this, the differentiator is usually preceded by a hard limiter that produces a square wave, and then by a bandpass filter centered at ω_c to change the signal back into sinusoidal form. A block diagram of the discriminator is given in Fig. 2.5. In practice, the differentiation operation can be achieved by designing a filter such that ω_c falls on a linearly sloping region of the filter magnitude response. Although the development has emphasized FM demodulation, the technique is also valid for PM signals if the envelope detector is followed by an integrator.

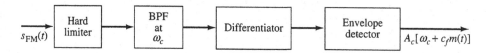

Fig. 2.5 FM discriminator

Other popular FM demodulators operate by detecting the zero crossings of the FM wave.

2.5   Phase-Locked Loops

For noisy channel applications, feedback demodulators are particularly powerful for recovering the message signal from the received FM waveform. Two types of feedback demodulators are the *FM feedback* (FMFB) *demodulator* and the *phase-locked loop* (PLL). Because of the increasing popularity of the PLL and its availability as an integrated-circuit device, we consider it in more detail here. In this section we investigate the basic operation of PLLs. More details on PLLs are left to the literature.

A block diagram of a phase-locked loop is shown in Fig. 2.6. The phase comparator in a PLL can be modeled as a multiplier followed by a unity-gain low-pass filter. As a result, $x(t)$ is given by

$$x(t) = \frac{A_c A_f}{2} \sin[(\omega_c - \omega_f)t + \phi_i(t) - \phi_f(t)], \qquad (2.5.1)$$

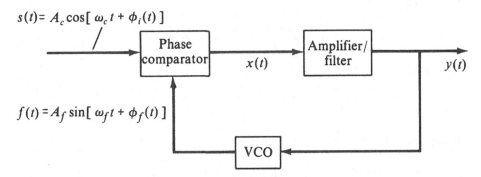

Fig. 2.6 Phase-locked loop

where the $\omega_c + \omega_f$ term is rejected by the LPF. Generally, we will ignore the frequency acquisition dynamics and assume that we have $\omega_c = \omega_f$ so

$$x(t) = \frac{A_c A_f}{2} \sin[\phi_i(t) - \phi_f(t)]. \qquad (2.5.2)$$

The VCO produces an output signal $f(t)$ with a frequency variation about ω_f that is proportional to its input signal. During ideal operation, we have $\phi_f(t) \cong \phi_i(t)$, so the frequency variation of $f(t)$ is

$$\frac{d}{dt}\phi_f(t) \cong \frac{d}{dt}\phi_i(t),$$

and since

$$\frac{d}{dt}\phi_f(t) = G_f y(t),$$

the PLL output is

$$y(t) = \frac{1}{G_f}\frac{d}{dt}\phi_i(t). \qquad (2.5.3)$$

Since $d\phi_i(t)/dt$ is the PLL input signal frequency variation, and since for FM and a message signal m(t) we know that

$$\frac{d}{dt}\phi_i(t) = c_f m(t), \tag{2.5.4}$$

it is evident that we have demodulated the FM waveform s(t).

When $\phi_f(t) \cong \phi_i(t)$ the PLL is said to be operating in *phase lock* with the input. The question remains as to how the PLL in Fig. 2.6 acquires and maintains this desirable situation. To begin, we return to Eq. (2.5.2) and note that if the gain of the amplifier/filter in Fig. 2.6 is G, then

$$y(t) = \frac{GA_cA_f}{2} \sin[\phi_i(t) - \phi_f(t)]. \tag{2.5.5}$$

Therefore,

$$\frac{d}{dt}\phi_f(t) = \frac{G_fGA_cA_f}{2} \sin[\phi_i(t) - \phi_f(t)], \tag{2.5.6}$$

so letting $G_T = (G_fGA_cA_f)/2$,

$$\phi_f(t) = G_T \int_{t_0}^{t_1} \sin[\phi_i(\tau) - \phi_f(\tau)]d\tau. \tag{2.5.7}$$

Defining the phase error $\theta_e(t) = \phi_i(t) - \phi_f(t)$, we then obtain

$$\frac{d}{dt}\theta_e(t) = \frac{d}{dt}\phi_i(t) - \frac{d}{dt}\phi_f(t). \tag{2.5.8}$$

If we assume that the PLL input signal has an abrupt change in frequency variation equal to $\Delta\omega_c = d\phi_i(t)/dt$, then we can write from Eq. (2.5.8) that

$$\frac{d}{dt}\phi_f(t) = \Delta\omega_c - \frac{d}{dt}\theta_e(t) = G_T \sin\theta_e(t), \tag{2.5.9}$$

where the last equality follows from Eq. (2.5.6). Equation (2.5.9) can be rewritten as a differential equation in the phase error as

$$\frac{d}{dt}\theta_e(t) + G_T \sin\theta_e(t) = \Delta\omega_c. \tag{2.5.10}$$

Equation (2.5.10) is sketched in Fig. 2.7.

The desired operating point in Fig. 2.7 is point O, since at that point the input frequency variation and the frequency variation of f(t) are equal $[d\theta_e(t)/dt = 0]$. Note that for any point on the curve in the upper half plane we have $d\theta_e(t) > 0$ for any positive change in time $dt > 0$. Hence any operating point in the upper half plane must move along the curve in the direction of increasing $\theta_e(t)$. Similarly, for any positive time increment, any operating point in the lower half plane has $d\theta_e(t)$ negative, so this operating point must move along the curve in the negative $\theta_e(t)$ direction. We therefore conclude that the point on the curve labeled "O" in Fig. 2.7 is the stable operating point of the PLL. The displacement of O from the origin along the $\theta_e(t)$ axis is the steady- state phase error $(\theta_e)_{ss}$. To make $(\theta_e)_{ss}$ small, the PLL parameters must be chosen appropriately. Further analyses of the PLL are relegated to the problems.

In summary, we see that the PLL operates to force $\theta_e(t) = \phi_i(t) - \phi_f(t) = (\theta_e)_{ss}$, and that for a well-designed loop, $(\theta_e)_{ss}$ will be small.

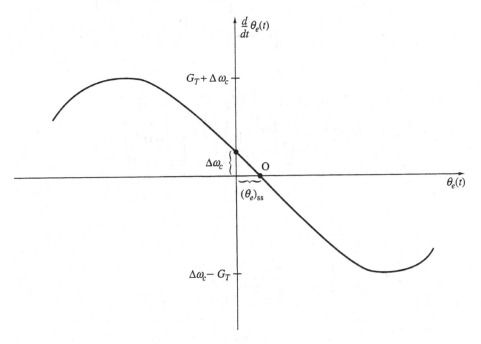

Fig. 2.7 Plot of Eq. (2.5.10)

2.6   Frequency Shift Keying

When frequency modulation is employed to transmit digital messages, particularly binary sequences, the modulation technique is called *frequency shift keying* (FSK). For a binary sequence, FSK simply consists of transmitting a single- frequency sinusoidal pulse for a logic 1 and a different frequency sinusoidal pulse for a logic 0. The pulse shape may vary depending on the application, but if we limit consideration in this section to rectangular pulses of width T, we have for the transmitted signal

$$s(t) = \begin{cases} A_c \cos \omega_1 t, & \text{for a } 1, \\ A_c \cos \omega_2 t, & \text{for a } 0, \end{cases} \qquad (2.6.1)$$

To determine the bandwidth requirements of an FSK signal, we need to model the series of 1's and 0's to be transmitted as a random binary sequence. We present such results shortly. However, to get a rough idea of the bandwidth occupied, we can investigate the alternating infinite sequence of 1's and 0's shown in Fig. 2.8a.

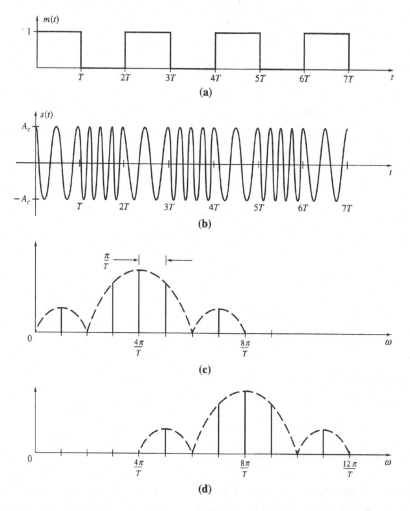

Fig. 2.8 Example 2.6.1, FSK: **a** Message sequence for Example 2.6.1; **b** FSK waveform for $m(t)$ in part **a**; **c** magnitude spectrum of $\cos \omega_1 t$ sequence; **d** magnitude spectrum of $\cos \omega_2 t$ sequence; **e** magnitude spectrum of the FSK wave

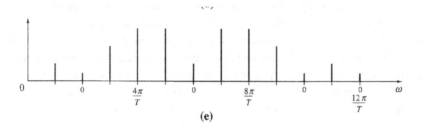

(e)

Fig. 2.8 (continued)

Example 2.6.1 If we let a pulse of amplitude A denote a 1 and no pulse a 0, the transmitted FSK waveform can be sketched as in Fig. 2.8b, where we have assumed that $\omega_1 = 2\pi/T_1 = 4\pi/T$ and $\omega_2 = 2\pi/T_2 = 8\pi/T$ in Eq. (2.6.1). To evaluate the required bandwidth, we note that the FSK waveform in Fig. 2.8b is simply the sum of two OOK sequences. The magnitude spectrum corresponding to the $\cos \omega_1 t$ carrier is easily calculated to be as shown in Fig. 2.8c, and the magnitude spectrum for the $\cos \omega_2 t$ sequence is given in Fig. 2.8d. These two spectra are summed to obtain the magnitude spectrum of the FSK signal sketched in Fig. 2.8e. Depending on which spectral lines we consider significant, we can state the required bandwidth for the FSK waveform.

We can also attack this problem by determining the maximum deviation of the instantaneous frequency. To do this, we assume that the carrier frequency $\omega_c = 6\pi/T$, so that the instantaneous frequency is $\omega_i = \omega_c \pm \Delta\omega = 6\pi/T \pm 2\pi/T$, and thus $\Delta\omega = 2\pi/T$. For Carson's rule we still need the message signal bandwidth. Since in general, $\omega_i = \omega_c + c_f m(t)$, we see that the normalized message signal is a sequence of alternating $+1$ and -1 amplitude rectangular pulses, each of width T. Writing a Fourier series for this sequence and assuming that the highest frequency of importance is at the second $\sin x/x$ zero crossing, we have that $\omega_m = 4\pi/T$.

Using Carson's rule, we find that

$$\text{BW} \cong 2(\omega_m + \Delta\omega) = 2\left(\frac{4\pi}{T} + \frac{2\pi}{T}\right) = \frac{12\pi}{T}, \tag{2.6.2}$$

which agrees (under our assumption of significant spectral lines) with Fig. 2.8e.

The exact calculation of the power spectral density of an FSK signal for a random sequence of 1's and 0's is relatively complicated, and hence the derivation is not given in this book. However, plots of the resulting spectral densities for various values of the frequency deviation are presented in Fig. 2.9. Note that as the frequency deviation nears $\Delta f = 1/2T$, a sharp peak appears in the spectral density at $f = 1/2T$ hertz. The situation in Example 2.6.1 has $\Delta\omega = 2\pi/T$, which corresponds to $h = 2$. The largest value of h shown in Fig. 2.9 is $h = 1.5$ in (d), but the similarity of the spectral shape to the spectral envelope in Fig. 2.8e is clear. For more details of the power spectral density of FSK, the reader is referred to Anderson and Salz (1965) or Proakis (1989).

Binary FSK signals can be demodulated using synchronous detection or noncoherent detection. We introduce the noncoherent approach here. Noncoherent detection of FSK is particularly popular in data transmission applications. A block diagram for noncoherent detection of FSK is shown in Fig. 2.10. The bandpass filters have a narrow passband, so that complete rejection of undesirable frequencies can be achieved.

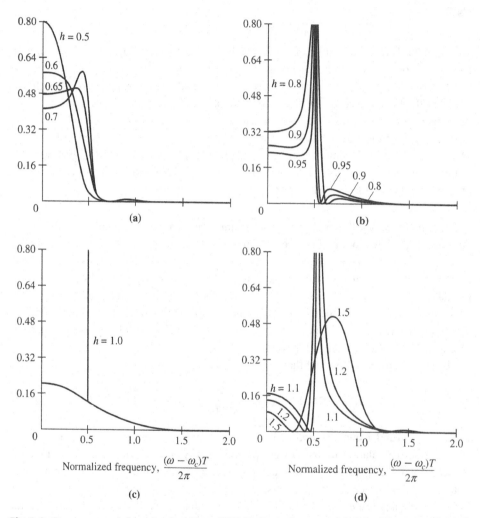

Fig. 2.9 Power spectral densities for binary FSK. R. R. Anderson and J. Salz, "Spectra of Digital FM," *Bell Systems Technical Journal,* Vol. 44, pp. 1165–1189, July-Aug., 1965. Copyright © 1965 AT&T. Reprinted by special permission

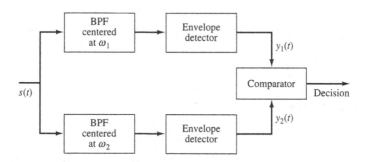

Fig. 2.10 Noncoherent detection of FSK

When $s(t) = A_c \cos \omega_1(t)$, $y_1(t)$ is a rectangular pulse and $y_2(t)$ is zero, so the comparator output is a logic 1. Similarly, when $s(t) = A_c \cos \omega_2 t$, the comparator "decides" that a logic 0 is present. The analysis of this detector in the presence of noise is not included here.

2.7 Phase Shift Keying

Phase shift keying (PSK) involves transmitting digital information by shifting the phase of a carrier among several discrete values. When a binary sequence is to be transmitted, the phase is usually switched between 0° and 180°, and the PSK signal is sometimes designated as *phase reversal keying* (PRK). Thus, for PRK, the transmitted signal $s(t)$ is

$$s(t) = \begin{cases} A_c \cos \omega_c t, & \text{for a logic 1} \\ -A_c \cos \omega_c t, & \text{for a logic 0.} \end{cases} \qquad (2.7.1)$$

Since for any chosen pulse shaping the envelopes of the two possible transmitted waveforms are the same, and further, they both have the same instantaneous frequency, we are forced to use coherent demodulation to recover the phase information and hence the message.

A block diagram of a detector for PRK is shown in Fig. 2.11. The filter $H(\omega)$ can be chosen to provide improved noise immunity in addition to rejecting the double-frequency terms. The remaining analysis of Fig. 2.11 is left to the reader.

The frequency content of a binary PSK(BPSK) waveform can be obtained using the frequency shifting property of the Fourier transform. In particular, for rectangular pulses of width τ = symbol interval, q = probability of a 1, and $1 - q$ = probability of a 0, the power spectral density of the baseband sequence is

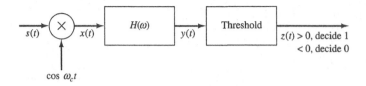

Fig. 2.11 Detector for PRK

$$S(\omega) = \frac{4q(1-q)}{\tau} \left| \frac{\tau \sin(\omega\tau/2)}{(\omega\tau/2)} \right|^2 + 2\pi (2q-1)^2 \delta(\omega). \qquad (2.7.2)$$

Note that if the binary symbols are equally likely $\left(q = 1 - q = \frac{1}{2}\right)$, the spectral line at $\omega = 0$ disappears and the spectral content is the magnitude squared of the familiar sin x/x shape. To obtain the frequency content of BPSK, $S(\omega)$ in Eq. (2.7.2) is shifted to $\pm\omega_c$.

Since PSK requires coherent demodulation, it is more common to see multiple-phase PSK than PRK. In four-phase PSK, pairs of binary digits are stored and each *pair* is represented by a different transmitted phase. Since there are four possible pairs of binary digits, we must have four different transmitted phases. The phases are usually spaced equally so that the transmitted phases for four-phase PSK are 90° apart. Any set of phases with this spacing will do, although one common choice is the set $\{\pm45°, +135°\}$.

The transmitted signal $s(t)$ is

$$s(t) = A_c \cos[\omega_c t + \theta_i], \qquad (2.7.3)$$

$i = 1, 2, 3, 4$, where we might assign the phases to pairs of binary digits, sometimes called *dibits*, as in Table 2.1. The transmitted signal in Eq. (2.7.3) can be rewritten using a trigonometric identity as

$$s(t) = A_c \cos\theta_i \cos\omega_c t - A_c \sin\theta_i \sin\omega_c t. \qquad (2.7.4)$$

In Eq. (2.7.4) we can clearly distinguish the in-phase and quadrature components, and from this equation we can surmise a possible implementation for the system transmitter.

Table 2.1 Four-phase PSK message sequence assignment

Dibit	Carrier phase (°)
00	+45
01	−45
10	+135
11	−135

Following Fig. 2.12 and using the phase assignments in Table 2.1, we see that for an input binary sequence of 01001110, we would first transmit

$$s(t) = \frac{A_c}{\sqrt{2}}[\cos \omega_c t + \sin \omega_c t], \qquad (2.7.5)$$

followed by

$$s(t) = \frac{A_c}{\sqrt{2}}[\cos \omega_c t - \sin \omega_c t]. \qquad (2.7.6)$$

The reader should complete the sequence.

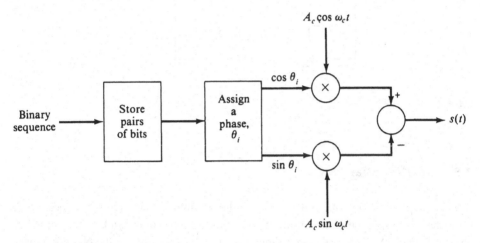

Fig. 2.12 Four-phase PSK transmitter

A demodulator for the transmitter in Fig. 2.12 would take the form shown in Fig. 2.13. For the phase assignment in Table 2.1, we obtain the decoding rule in Table 2.2. It is important to notice that an erroneous phase reference in the local oscillator can cause decoding errors in PSK, and hence a PLL, a transmitted pilot tone, or some other technique must be used to guard against such an eventuality.

Table 2.2 Decoding rule for Fig. 2.13 and the phase assignment in Table 2.1

f_i	$-f_q$	θ_i (°)	Decoded dibit
$\frac{A_c}{\sqrt{2}}$	$\frac{-A_c}{\sqrt{2}}$	45	00
$\frac{A_c}{\sqrt{2}}$	$\frac{A_c}{\sqrt{2}}$	−45	01
$\frac{-A_c}{\sqrt{2}}$	$\frac{-A_c}{\sqrt{2}}$	+135	10
$\frac{-A_c}{\sqrt{2}}$	$\frac{A_c}{\sqrt{2}}$	−135	11

It is noted that the coherent demodulator in Fig. 2.13 is not optimum in terms of minimizing the probability of a bit error.

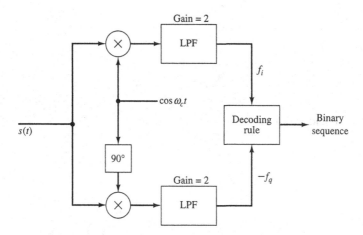

Fig. 2.13 Demodulator for four-phase PSK

2.8   Differential PSK

Differential phase shift keying (DPSK) avoids the requirement for an accurate local oscillator phase by using the phase during the immediately preceding symbol interval as the phase reference. As long as the preceding phase is received correctly, the phase reference is accurate. Therefore, this approach presupposes that the channel will not produce a serious enough phase error in one symbol interval to cause an erroneous detected phase at the receiver.

There are a variety of encoding techniques for implementing a DPSK system. One technique, which can be used for one-bit-at-a-time transmission, is to obtain a differential binary sequence from the input binary sequence and then assign phases to the bits in the differential sequence. This encoding technique is illustrated in Table 2.3. The reference bit is preselected and fixed, but otherwise it is arbitrary. The differential binary sequence is generated by repeating the preceding bit in the differential sequence if the message bit is a 1 or by changing to the opposite bit if the message bit is a 0. Phases are then assigned to the differential binary sequence by transmitting 0° for a 1 and 180° for a 0. The entire procedure is demonstrated in Table 2.3.

Table 2.3 One-bit-at-a-time DPSK scheme

Binary message sequence

	0	1	1	0	1	0	0	1	1	1

Differential binary sequence

1	0	0	0	1	1	0	1	1	1	1

Reference bit

Transmitted phase

0°	180°	180°	180°	0°	0°	180°	0°	0°	0°	0°

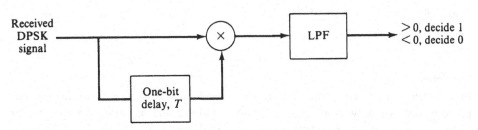

Fig. 2.14 DPSK demodulator for the encoding in Table 2.3

Recovery of the binary message sequence can also be accomplished in several ways. One method is illustrated in Fig. 2.14. The received DPSK signal is $\pm\cos\omega_c t$. If the current signal and the immediately preceding received signal agree, the LPF output is positive, and it is decided that the message sequence contains a 1. If the received DPSK signal and its predecessor differ, the LPF output is negative, and the message bit is a 0.

Another common DPSK encoding scheme is to group bits into singles, pairs (dibits), or triples (tribits), and then associate a particular phase *change* with each group. As an example, consider the situation shown in Table 2.4, where each pair of binary digits or dibit is represented by the change in transmitted phase indicated. If we wish to transmit the binary message sequence shown in Table 2.5, and we assume an initial reference phase of 0°, the transmitted phases are as indicated in the table. Note that the difference between the preceding phase and the present transmitted phase corresponds to the appropriate dibit. Demodulation is accomplished by detecting the received phase sequence, comparing adjacent phases, and outputting the corresponding dibit.

It is common today to use DPSK to transmit tribits, but larger groups of bits are prohibited by phase errors. For many years, DPSK has proven to be a highly effective method for data transmission due to its relative simplicity and reliability.

Table 2.4 Dibits and transmitted phase changes

Dibit	Phase change (°)
00	45
01	135
10	225
11	315

Table 2.5 Four-phase DPSK example

Binary message sequence		1 0	1 1	0 0	1 0	1 0
Transmitted phase	0°	225°	180°	225°	90°	315°
Initial phase reference	↑					

Summary

In this chapter we have introduced the important techniques of frequency modulation and phase modulation as well as the concepts of wideband and narrowband FM and their bandwidth requirements. Standard modulation and demodulation methods were also presented, including an introduction to phase- locked loops. The use of phase and frequency modulation for data transmission was also considered, with discussions and developments of FSK, PSK, and DPSK.

Problems

2.1 Given the waveform

$$s(t) = 5\cos\left[5 \times 10^6 t + 100\sin 200t\right],$$

  (a) If $s(t)$ is a PM wave, what is the message signal $m(t)$?
  (b) If $s(t)$ is an FM signal, what is $m(t)$?

2.2 A transmitted angle-modulated waveform is given by.

$$s(t) = 20\cos\left[10^6 t + 10\cos 500t\right].$$

  (a) What is the instantaneous frequency of $s(t)$?
  (b) What is the approximate bandwidth of $s(t)$?
  (c) If $s(t)$ is a PM wave, what is the message signal $m(t)$?
  (d) If $s(t)$ is an FM signal, what is the message signal $m(t)$?

2.3 For a message signal given by $m(t) = A_m\cos\omega_m t$, write expressions for the modulation index if:

(a) $m(t)$ is sent by FM.

(b) $m(t)$ is sent by PM. In which case is β independent of ω_m?

2.4 The message $m(t)$ sketched in Fig. 2.15 is to be transmitted using both PM and FM.

(a) For PM let $c_p = \pi/2$ rad/V and assume that $T = 3T_c$, where the carrier frequency $\omega_c = 2\pi/T_c$. Sketch the PM wave.

(b) For FM let $c_f = \pi/T$ rad/s per volt, with $T = 3T_c$ and $\omega_c = 2\pi/T_c$. Sketch

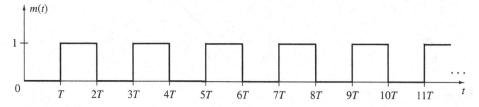

Fig. 2.15 Message Waveform for Problem 2.4

the FM wave.

2.5 Consider the message signal shown in Fig. 2.16.

(a) If $m(t)$ is to be transmitted using FM, let the carrier frequency $\omega_c = 2\pi/T_c$, where $T_c = T/4$ and $c_f = \pi/T$ rad/s per volt, and sketch the transmitted waveform.

(b) If $m(t)$ is to be transmitted using PM, let $\omega_c = 2\pi/T_c$, $T_c = T/4$, and $c_p = \pi/2$rad/V, and sketch the PM waveform.

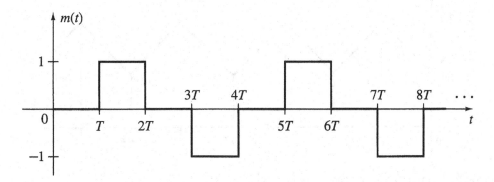

Fig. 2.16 Message Waveform for Problem 2.5

2.6 If $\theta(t)$ in Eq. (2.2.1) has the form shown in Fig. 2.17, sketch $s_{PM}(t)$. Let $\omega_c = 2\pi/T_c = 8\pi/T$.

2.7 If the instantaneous frequency of an FM wave is as shown in Fig. 2.18, sketch the FM waveform. Let $\omega_c = 2\pi/(T/2)$ and $A_c = 1$..

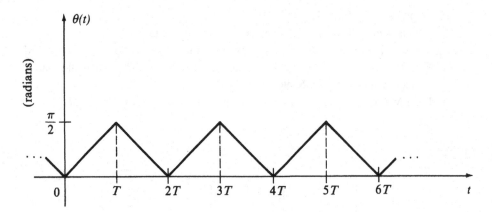

Fig. 2.17 Phase for Problem 2.6

2.8 A 10-MHz carrier signal is frequency modulated by a unit amplitude sinusoid with $\omega_m = 2000\pi$ rad/s. If $c_f = 10$ rad/s per volt:
   (a) What is the modulation index?
   (b) Is this a wideband or a narrowband signal? Why?
   (c) What is the bandwidth of the transmitted signal?

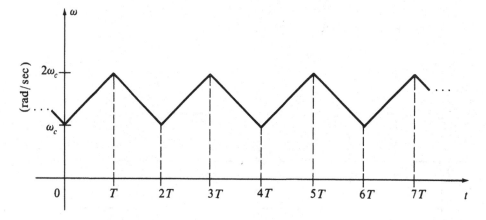

Fig. 2.18 Instantaneous Frequency for Problem 2.7

2.9 Given the angle-modulated waveform

$$s(t) = 10\cos\left[90 \times 10^6 t + 200\sin 4000\pi t\right].$$

   (a) What is the modulation index?
   (b) Is $s(t)$ wideband or narrowband? Why?

(c) What is the required bandwidth for $s(t)$?

2.10 Consider a message signal consisting of the sum of two sinusoids,

$$m(t) = A_1 \cos \omega_{m1} t + A_2 \cos \omega_{m2} t.$$

If $m(t)$ is to be transmitted via FM:

(a) Write an expression for the transmitted FM waveform.

(b) What is the maximum frequency deviation?

(c) Obtain an expression for the FM waveform that we can use to determine its bandwidth requirements.

2.11 Show that angle modulation is nonlinear.

2.12 The average power in an FM wave that communicates a pure tone sinusoidal message is $\langle s^2(t) \rangle = (A_c^2/2) \sum_{n=-\infty}^{\infty} J_n^2(\beta)$. For maximum efficiency, we wish all of the power to be in the sidebands. Show how 100% efficiency in this FM signal can be achieved.

2.13 The average power in an angle-modulated signal is $\langle s^2(t) \rangle = A_c^2/2$. Show that $\sum_{n=-\infty}^{\infty} J_n^2(\beta) = 1$.

2.14 Consider an FM wave modulated by a single-tone sinusoid as in Eq. (2.3.4). Sketch the magnitude spectrum of $s_{FM}(t)$ as the amplitude of the modulating signal is increased but ω_m is held fixed. Specifically consider the cases $\beta = 0.1$, $\beta = 1$, and $\beta = 10$.

2.15 For an FM wave modulated by a single-tone sinusoid as in Eq. (2.3.4), sketch the magnitude spectrum of $s_{FM}(t)$ if $\Delta\omega = 60\pi$ rad/s is held fixed but ω_m is varied. Specifically, let $\omega_m = 60\pi, 6\pi$ and 2π rad/s.

2.16 Plot the bandwidth given by Carson's rule versus β when

(a) The bandwidth is normalized by the frequency deviation $\Delta\omega$.

(b) The bandwidth is normalized by the message signal bandwidth ω_m.

2.17 Retaining only those sidebands with an amplitude that is 10% of the unmodulated carrier amplitude or greater, plot the bandwidth of the single-tone modulated FM wave versus β when:

(a) The bandwidth is normalized to $\Delta\omega$.

(b) The bandwidth is normalized to ω_m.

2.18 Repeat Problem 2.17 for sidebands with an amplitude that is 1% of the unmodulated carrier amplitude.

2.19 A narrowband FM wave is given by

$$x(t) = A_c \cos\left[2\pi \times 10^5 t\right] - 0.01 A_c \sin[2000\pi t] \sin\left[2\pi \times 10^5 t\right].$$

(a) If we write $x(t) = A_c \cos[\omega_c t + \psi(t)]$, what is $\psi(t)$?

(b) Draw a block diagram of an indirect FM transmitter that will generate

$$s(t) = A_c \cos\left[2\pi \times 10^6 t + \sin 2000\pi t\right]$$

from $x(t)$.

2.20 A transmitted angle modulated signal is given by $s(t) = A_c \cos[\omega_c t + \theta(t)]$. Unfortunately, an interfering tone is present during transmission, so that the received signal is

$$r(t) = s(t) + A_d \cos[(\omega_c + \omega_d)t].$$

(a) Find an expression for the envelope of $r(t)$.
(b) Find an expression for the phase of $r(t)$.
(c) If $A_c \gg A_d$, what is the approximate instantaneous frequency of $r(t)$?
(d) If $A_d \gg A_c$, what is the approximate instantaneous frequency of $r(t)$?

2.21 A block diagram of a typical broadcast FM receiver is shown in Fig. 2.19. If the IF amplifier center frequency is 10.7 MHz and the IF band-width is 200 kHz, is there a problem with image stations in broadcast FM?

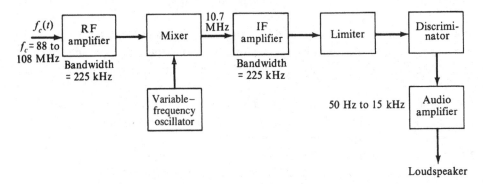

Fig. 2.19 FM Receiver for Problem 2.21

2.22 For stereo FM, we multiplex the two message signals, along with a pilot tone, and then use narrowband FM. The multiplexing operation is indicated in Fig. 2.20.
(a) If $m_1(t)$ and $m_2(t)$ have the Fourier transforms shown, sketch the Fourier transform of $s(t)$ in Fig. 2.20.
(b) If we use narrowband FM, what is the bandwidth required to transmit $s(t)$?

2.23 A block diagram of an FM stereo receiver is shown in Fig. 2.21a. If the discriminator output has the Fourier transform shown in Fig. 2.21b, verify that the LPF outputs are $m_1(t)$ and $m_2(t)$. Note that for monaural FM receivers to be compatible with stereo FM transmission, $m_1(t)$ must represent the sum of the left and right stereo channels. If we then make $m_2(t)$ equal the difference between the left and right

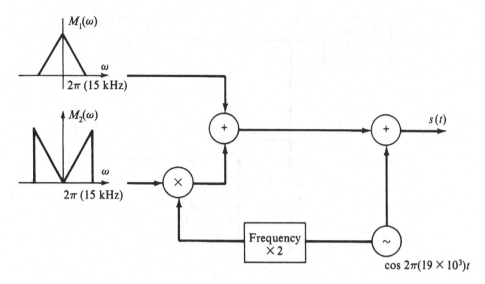

Fig. 2.20 Stereo FM Multiplexing for Problem 2.22

channels, we can obtain the left and right channels separately in a stereo receiver by adding and subtracting $m_1(t)$ and $m_2(t)$.

2.24 A nonlinear channel with input $x(t)$ and output $y(t)$ has the following input/output relationship:

$$y(t) = c_1 x(t) + c_2 x^2(t) + c_3 x^3(t).$$

If $x(t) = s_{FM}(t) = A_c \cos\left[\omega_c t + c_f \int m(t)dt\right]$, show how $s_{FM}(t)$ can be recovered undistorted at the receiver. Compare AMDSB-SC in Problem 1.9.

2.25 Show that any square-law device with input/output relationship $y(t) = x^2(t)$ can be used as a frequency multiplier.

2.26 Given the narrowband angle-modulated waveform

$$s(t) = A_c \cos \omega_c t - A_c g(t) \sin \omega_c t,$$

where $g(t) = c_p m(t)$ for PM or $g(t) = c_f \int m(t)dt$ for FM:

(a) Write $s(t)$ in the envelope-phase form, and thus show that $s(t)$ has amplitude and phase distortion.

(b) Find an infinite series expansion for the instantaneous frequency.

(c) Let $m(t) = A_m \cos \omega_m t$ and assume FM transmission. Retaining only third harmonics and below ($\beta < 1$), write an expression for the instantaneous frequency.

2.27 Use Eq. (2.5.8) to obtain the first-order differential equation for the PLL,

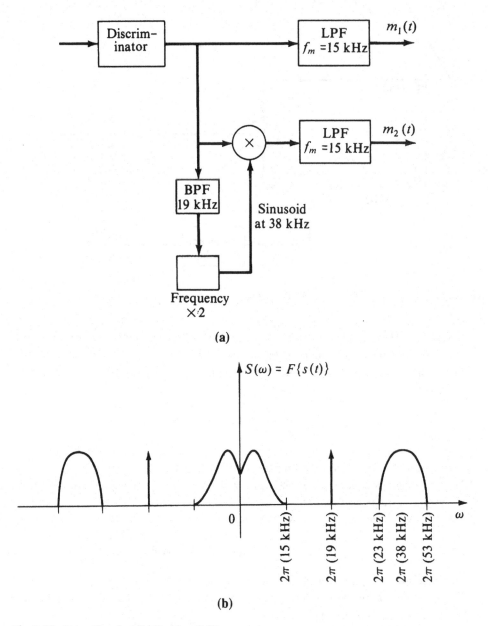

(a)

(b)

Fig. 2.21   Stereo Receiver for Problem 2.23

$$\frac{d}{dt}\theta_e(t) + G_T\theta_e(t) = \frac{d}{dt}\phi_i(t).$$

Hint: Assume that $\theta_e(t)$ is small.

2.28 Letting $r(t) = d\phi_i(t)/dt$, use Laplace transforms to obtain the transfer function of a first-order PLL, $\theta_e(s)/R(s)$, from Problem 2.27.

(a) For a step change in frequency, $R(s) = 1/s$. Show that the steady-state value of $\theta_e(t)$ is $(\theta_e)_{ss} = 1/G_T$.

(b) What is the 3-dB bandwidth of the first-order loop?

(c) Discuss the results of parts (a) and (b).

2.29 Perform the required calculations and sketch the spectra for Example 2.6.1 if $\omega_1 = 20\pi/T$ and $\omega_2 = 24\pi/T$.

2.30 The PLL analysis in Sect. 2.5 assumes that the amplifier/filter inside the loop in Fig. 2.6 has a unity transfer function for all frequencies. Let the transfer function of this filter be $H(\omega)$ and find an expression for $d\theta_e(t)/dt$.

2.31

(a) Use the small-angle approximation on Eq. (2.5.6) to obtain a linear model of the PLL.

(b) Repeat part (a) for Problem 2.30.

(c) Find the transfer function of part (b), $\Phi_f(\omega)/\Phi_i(\omega)$.

2.32 Show that an RC high-pass filter can be used to obtain Eq. (2.4.7).

Hint: Assume for the frequencies of interest that $\omega \ll 1/RC$.

2.33 Find the transfer function of the first-order, linearized PLL from $\Phi_i(s) = \mathcal{L}\{\phi_i(t)\}$ to $\theta_e(s)$. For a step change in input frequency, find an expression for $\theta_e(t)$.

2.34 For a second-order PLL, the amplifier/filter inside the loop in Fig. 2.6 has (approximately)

$$H(s) = \frac{1 + s/\alpha}{s/\beta}.$$

(a) Find the transfer function $\Phi_f(s)/\Phi_i(s)$ of a linearized, second-order PLL.

(b) For a step change in input frequency, $\Phi_i(s) = \Delta\omega/s^2$, find $\theta_e(t)$.

(c) Let $t \to \infty$ in $\theta_e(t)$ and compare to a first-order PLL.

2.35 Can the approach represented by the diagram in Fig. 2.22 be used for FM demodulation? Under what conditions?

2.36 Eight-phase PSK is used to transmit binary data sequences by assigning messages (tribits) to phases as shown in Table 2.6. Assume that the carrier frequency is ω_c.

(a) If the sequence to be transmitted is 1010010001 1 1, write expressions for the transmitted waveforms in terms of in-phase and quadrature components.

(b) Specify a decoding table relating the received in-phase and quadrature components and the decoded bits.

2.37 A one-bit-at-a-time DPSK encoding scheme does not generate a differential binary sequence but simply transmits no phase change for a binary message 1 and a 180° phase change for a binary message 0. Assuming an initial phase reference of 0°, use

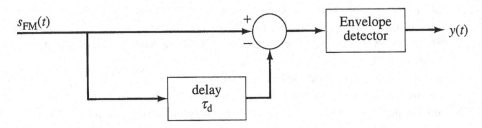

Fig. 2.22 Block Diagram for Problem 2.35

Table 2.6 Encoding rule for Problem 2.36

Tribit	000	001	010	011	100	101	110	111
Carrier phase	0°	45°	90°	135°	180°	−45°	−90°	−135°

this scheme to transmit the binary message sequence in Table 2.3. Compare your transmitted phases to the transmitted phase sequence in the table.

2.38 Tribits are to be transmitted using eight-phase DPSK. The tribits and their associated phase changes are listed in Table 2.7. Assuming a 0° initial phase reference, specify the transmitted phase sequence that represents the binary message sequence 1000001 1 1 10101000 1.

Table 2.7 Tribits and transmitted phase changes for Problem 2.38

Tribit	000	001	010	011	100	101	110	111
Phase change	22.5°	67.5°	112.5°	157.5°	202.5°	247.5°	292.5°	337.5°

Noise in Analog Modulation

3

3.1   Introduction

When designing a communication system, the electrical engineer is usually confronted with limitations on transmitted power and available bandwidth. Required power and bandwidth 'figured prominently in our discussions of analog modulation techniques, AM, FM, and PM, in Chaps. 1 and 2. In this chapter we consider the performance of these analog modulation methods in the presence of noise. Noise is an unavoidable aspect of any communication system. However, if we are willing (and able) to pay the price, say, in terms of excessive transmitted power or system complexity, the noise can be made to have a negligible impact on the performance of a communication system. In most instances, however, we are interested in designing the least expensive, most efficient, highest-performance system possible within the limits of our specifications. Of course, in these situations, analyses of the effects of noise on communication system performance are vitally important. Hence we are led to the purpose of this chapter.

In communication system analyses, there are deterministic impairments such as channels with nonflat amplitude responses and nonlinear phase characteristics, and these deterministic distortions can have a significant impact on communication system performance. However, our interest in this chapter is in random impairments, that is, in distortions that can be modeled as being generated by some random experiment. We call this random distortion *noise*. We also assume throughout the analyses in this chapter that the noise is additive, so that the received signal after transmission over the channel is the undistorted modulated signal plus a random noise waveform. Additive noise is common in practical communication systems, and hence this is a natural place to begin our investigations.

In Sect. 3.2 the relevant noise models are developed, and in the following sections we use these noise models to obtain objective measures of the various AM, FM, and PM

© The Author(s), under exclusive license to Springer Nature Switzerland AG 2023
J. D. Gibson, *Analog Communications*, Synthesis Lectures on Communications,
https://doi.org/10.1007/978-3-031-19584-6_3

system performances. The principal objective performance indicator is the system output signal-to-noise ratio (SNR). At the end of the chapter we are able to compare the various analog modulation systems on the basis of signal-to-noise ratios, transmitted power, and bandwidth.

3.2   Narrowband Noise

To begin our discussion, we consider a general time function, say $x(t)$, with Fourier transform $X(\omega)$. If $X(\omega)$ is nonzero only over a band of frequencies given by $\omega_0 - \omega_B \leq |\omega| \leq \omega_0 + \omega_B$, where $\omega_0 \gg \omega_B$, $x(t)$ is called a *narrowband waveform* and has a magnitude spectrum of the form shown in Fig. 3.1. From our knowledge of amplitude modulation, we see that the spectrum in Fig. 3.1 is reminiscent of a carrier wave with a slowly varying envelope, or possibly, from our developments of frequency and phase modulation, a carrier with a slowly varying frequency or phase (narrowband FM). It is thus plausible that $x(t)$ could be expressed as

$$x(t) = \rho(t) \cos[\omega_c t + \theta(t)], \tag{3.2.1}$$

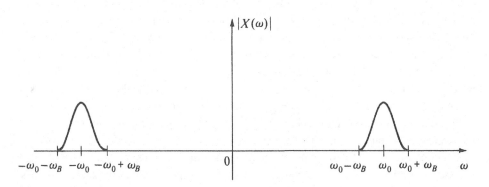

Fig. 3.1 Magnitude spectrum of a narrowband waveform

where $\rho(t)$ is the slowly varying envelope and $\theta(t)$ is the slowly varying phase. We know that we can use trigonometric identities to expand $x(t)$ as

$$x(t) = \rho(t) \cos \theta(t) \cos \omega_c t - \rho(t) \sin \theta(t) \sin \omega_c t$$
$$= x_i(t) \cos \omega_c t - x_q(t) \sin \omega_c t, \tag{3.2.2}$$

where

$$x_i(t) \triangleq \rho(t) \cos \theta(t) \tag{3.2.3}$$

$$x_q(t) \triangleq \rho(t) \sin \theta(t) \tag{3.2.4}$$

are the in-phase and quadrature components of the signal, respectively. Upon using Eqs. (3.2.3) and (3.2.4), it is immediate that the envelope and phase can be expressed in terms of the in-phase and quadrature components as

$$\rho(t) = \sqrt{x_i^2(t) + x_q^2(t)} \tag{3.2.5}$$

and

$$\theta(t) = \tan^{-1} \frac{x_q(t)}{x_i(t)}. \tag{3.2.6}$$

Now, our interest in this section is noise, and it is common to model the random disturbances contributed by the channel in communication systems as an additive white Gaussian noise process $\{X(t), -\infty < t < \infty\}$ with zero mean and two-sided spectral density $S_X(\omega) = (2\pi)\mathcal{N}_0/2$ watts/rad per second for $-\infty < \omega < \infty$. Clearly, this is a wideband process, and we wonder how narrowband noise is obtained from $\{X(t)\}$. If one examines the analog communication systems in Chaps. 1 and 2, the answer becomes quite clear. At the front end of the receivers in analog communication systems, the received signals are passed through at least one bandpass filter whose bandwidth is proportional to the message signal frequency content, say ω_B. Since $\omega_B \ll \omega_c$, the noise process at the output of these filters, denoted by $\{N(t), -\infty < t < \infty\}$, is narrowband with a spectral density $S_N(\omega) = \pi|H(\omega)|^2\mathcal{N}_0$ watts/rad per second, where $H(\omega)$ is the receiver bandpass filter transfer function. The noise spectral content at the bandpass filter output thus greatly resembles the shape shown in Fig. 3.1, and therefore the noise process $\{N(t)\}$ can be written in narrowband form. At this point, to conform with the existing literature and to simplify the notation in subsequent sections, we make a change of notation and let the narrowband noise process be represented by $\{n(t), -\infty < t < \infty\}$. rather than $\{N(t), -\infty < t < \infty\}$. This will be used henceforth for narrowband noise processes.

With this change in notation, we can now write the narrowband noise process $\{n(t)\}$ as

$$n(t) = \rho(t) \cos[\omega_c t + \theta(t)] = n_i(t) \cos \omega_c t - n_q(t) \sin \omega_c t, \tag{3.2.7}$$

where $\rho(t)$ and $\theta(t)$ are now random processes and

$$n_i(t) = \rho(t) \cos \theta(t) \tag{3.2.8}$$

and

$$n_q(t) = \rho(t) \sin \theta(t) \tag{3.2.9}$$

Again, the expressions for $\rho(t)$ and $\theta(t)$ follow as

$$\rho(t) = \sqrt{n_i^2(t) + n_q^2(t)} \qquad (3.2.10)$$

and

$$\theta(t) = \tan^{-1} \frac{n_q(t)}{n_i(t)}. \qquad (3.2.11)$$

Since $\{n(t)\}$ is obtained by linearly filtering the Gaussian process $\{X(t)\}$, $n(t)$ is Gaussian; and further, $n_i(t)$ and $n_q(t)$ can be derived from $n(t)$ by linear transformations, and hence they are also Gaussian processes. We assume for the sequel that $n(t)$ is zero mean.

It is often useful to have available the marginal pdfs of the envelope $\rho(t)$ and the phase $\theta(t)$ Since $n_i(t)$ and $n_q(t)$ are Gaussian processes, and if we assume that they are zero mean (see Property 1 later in this section), then it is straightforward to show via a transformation of variables that the joint pdf of $\rho(t)$ and $\theta(t)$ is

$$f_{\rho,\theta}(\rho, \theta; t) = \frac{\rho(t)}{2\pi\sigma^2} e^{-\rho^2(t)/2\sigma^2} \qquad (3.2.12)$$

for $0 \leq \rho(t) < \infty$ and $0 \leq \theta(t) \leq 2\pi$ The marginal pdfs follow directly as

$$f_\rho(\rho; t) = \int_0^{2\pi} f_{\rho,\theta}(\rho, \theta; t) d\theta(t)$$
$$= \frac{\rho(t)}{\sigma^2} e^{-\rho^2(t)/2\sigma^2} \qquad (3.2.13)$$

for $0 \leq \rho(t) < \infty$, and

$$f_\theta(\theta; t) = \int_0^\infty f_{\rho,\theta}(\rho, \theta; t) d\rho(t)$$
$$= \frac{1}{2\pi} \qquad (3.2.14)$$

for $0 \leq \theta(t) \leq 2\pi$. The pdf in Eq. (3.2.13) is called a *Rayleigh* probability density function, while Eq. (3.2.14) indicates that $\theta(t)$ is uniformly distributed over the interval $[0, 2\pi]$.

We now list several properties of narrowband Gaussian processes without proof, which will be useful in later sections.

Property 3.1 If $n(t)$ is zero mean, $n_i(t)$ and $n_q(t)$ are zero mean.

Property 3.2 Since the in-phase and quadrature components of a narrowband process are uncorrelated and since $n_i(t)$ and $n_q(t)$ are jointly Gaussian, then $n_i(t)$ and $n_q(t)$ are independent.

Property 3.3 The in-phase and quadrature components have the same autocorrelation function, and hence the same spectral density.

Property 3.4 If $n(t)$ is wide-sense stationary, the $n_i(t)$ and $n_q(t)$ are jointly wide-sense stationary.

We can find the mean-squared value of $n(t)$ starting with Eq. (3.2.7) as

$$E\{n^2(t)\} = E\{n_i^2(t)\} \cos^2 \omega_c t + E\{n_q^2(t)\} \sin^2 \omega_c t, \qquad (3.2.15)$$

where we have employed Properties 1 and 2 to eliminate the cross terms. Using trigonometric identities and invoking Property 3.3, we see that

$$E\{n^2(t)\} = E\{n_i^2(t)\} = E\{n_q^2(t)\}$$

$$= \frac{1}{2}E\{n_i^2(t)\} + \frac{1}{2}E\{n_q^2(t)\}. \qquad (3.2.16)$$

Additional developments concerning narrowband noise are included in the problems.

3.3   AM-Coherent Detection

For the purposes of evaluating the performance of the individual analog modulation systems and comparing the performances among systems, we must select an analytically tractable and physically meaningful quantity. For our developments in this section and the remainder of the chapter, we choose to examine the receiver output signal-to-noise ratio for a specified input signal-to-noise ratio, where we define the signal-to-noise ratio as the ratio of the time average of the squared "message" signal component to the mean-squared value of the noise at the same point. For all the developments, we assume that the channel has added white Gaussian noise which has been bandpass filtered to generate narrowband Gaussian noise at the detector inputs.

For AMDSB-SC signals, the receiver after bandpass filtering can be represented by the block diagram in Fig. 3.2. The received waveform after bandpass filtering is given by

$$r(t) = A_1 m(t) \cos \omega_c t + n_i(t) \cos \omega_c t - n_q(t) \sin \omega_c t \qquad (3.3.1)$$

Where $m(t)$ is the low-pass message signal bandlimited to ω_m. Upon considering *r(t)* to be the receiver input waveform, it is clear that $A_1 m(t) \cos \omega_c t$ is the desired signal component and the other terms in Eq. (3.3.1) constitute the bandpass Gaussian noise. Therefore, the average power in the desired component is

$$A_1^2\langle m^2(t) \cos^2 \omega_c t\rangle = \frac{A_1^2}{2}\langle m^2(t)\rangle, \qquad (3.3.2)$$

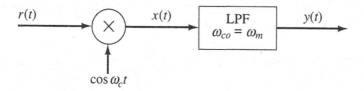

Fig. 3.2 Coherent detector for AMDSB-SC and AMSSB-SC

where $\langle \cdot \rangle$ denotes time averaging, and the mean-squared value of the noise is $E\{n^2(t)\}$ as given by Eq. (3.2.15). The input signal-to-noise ratio, SNR_i, is thus

$$SNR_i = \frac{(A_1^2/2)\langle m^2(t)\rangle}{E\{n^2(t)\}}.$$ (3.3.3)

To determine the output signal-to-noise ratio, SNR_o, we need an expression for $y(t)$, which follows easily from Fig. 3.2 as

$$y(t) = \frac{A_1}{2}m(t) + \frac{1}{2}n_i(t).$$ (3.3.4)

The message signal power and mean-squared value of the noise component are thus $A_1^2\langle m^2(t)\rangle/4$ and $E\{n_i^2(t)\}/4$, respectively, so that the output signal-to-noise ratio is

$$SNR_o = \frac{A_1^2\langle m^2(t)\rangle}{E\{n_i^2(t)\}} = \frac{A_1^2\langle m^2(t)\rangle}{E\{n^2(t)\}},$$ (3.3.5)

where the last equality follows from Eq. (3.2.16). Comparing Eqs. (3.3.3) and (3.3.5), we see that for AMDSB-SC,

$$SNR_o = 2SNR_i.$$ (3.3.6)

The coherent receiver thus provides a factor of 2 or a 3-dB gain in SNR by eliminating the quadrature component of the narrowband noise.

The coherent receiver block diagram in Fig. 3.2 is also valid for AMSSB-SC waveforms; however, in this case the received signal after bandpass filtering is

$$r(t) = A_2 m(t)\cos\omega_c t \pm A_2 m_h(t)\sin\omega_c t + n_i(t)\cos\omega_c t - n_q(t)\sin\omega_c t,$$ (3.3.7)

where $m_h(t)$ is the Hilbert transform of $m(t)$. The average power in the desired signal component is

$$A_2^2\langle[m(t)\cos\omega_c t \pm m_h(t)\sin\omega_c t]^2\rangle = \frac{A_2^2}{2}\langle m^2(t)\rangle + \frac{A_2^2}{2}\langle m_h^2(t)\rangle$$

$$= A_2^2\langle m^2(t)\rangle,$$ (3.3.8)

where the last equality results from property (c) in Problem 1.15. Check the input noise power is again just $E\{n^2(t)\}$, so

$$\text{SNR}_i = \frac{A_2^2\langle m^2(t)\rangle}{E\{n^2(t)\}}. \tag{3.3.9}$$

After synchronous detection, we have that

$$y(t) = \frac{A_2}{2}m(t) + \frac{1}{2}n_i(t), \tag{3.3.10}$$

so that the output signal-to-noise ratio follows immediately as

$$\text{SNR}_o = \frac{A_2^2\langle m^2(t)\rangle}{E\{n^2(t)\}}. \tag{3.3.11}$$

Comparing Eqs. (3.3.9) and (3.3.11), we get the result that

$$\text{SNR}_o = \text{SNR}_i \tag{3.3.12}$$

for coherent detection of AMSSB-SC. The receiver does not provide a gain in SNR for SSB as it does for DSB, since the desired signal power in the $m_h(t)$ component is discarded by the receiver along with the quadrature component of the noise.

3.4 AM-Noncoherent Detection

For AMDSB-TC or conventional AM, the receiver simply consists of an envelope detector, and the input to this detector after bandpass filtering is

$$r(t) = A_3[1 + am(t)]\cos\omega_c t + n_i(t)\cos\omega_c t - n_q(t)\sin\omega_c t, \tag{3.4.1}$$

where a is the modulation index and $m(t)$ is assumed to be normalized here. The only term in Eq. (3.4.1) containing the message signal is the component $aA_3m(t)\cos\omega_c t$, which has an average power

$$\langle a^2 A_3^2 m^2(t)\cos^2\omega_c t\rangle = \frac{1}{2}a^2 A_3^2\langle m^2(t)\rangle. \tag{3.4.2}$$

Since the mean-squared value of the noise is $E\{n^2(t)\}$, we have that the input SNR is

$$\text{SNR}_i = \frac{a^2 A_3^2\langle m^2(t)\rangle}{2E\{n^2(t)\}} \tag{3.4.3}$$

To find the output SNR, we need the envelope of $r(t)$ in Eq. (3.4.1). Although many alternative approaches are possible here, we choose to use Eqs. (3.2.8) and (3.2.9) and write

$$\text{env}[r(t)] = \left\{ A_3^2[1 + am(t)]^2 + 2A_3[1 + am(t)]\rho(t) \cos\theta(t) + \rho^2(t) \right\}^{1/2}. \qquad (3.4.4)$$

To simplify this, we factor $A_3^2[1 + am(t)]^2$ out of the radical and assume that we are in the high carrier-to-noise ratio case,

$$A_3^2 \gg E\{\rho^2(t)\},$$

so that

$$\text{env}[r(t)] = A_3[1 + am(t)] \left\{ 1 + \frac{2\rho(t) \cos\theta(t)}{A_3[1 + am(t)]} + \frac{\rho^2(t)}{A_3^2[1 + am(t)]^2} \right\}^{1/2}$$

$$\cong A_3[1 + am(t)] \left\{ 1 + \frac{2\rho(t) \cos\theta(t)}{A_3[1 + am(t)]} \right\}^{1/2}. \qquad (3.4.5)$$

Using the approximation for the square root that $\{1 + x\}^{1/2} \cong 1 + \frac{1}{2}x$, we have

$$\text{env}[r(t)] \cong A_3[1 + am(t)] + \rho(t) \cos\theta(t)$$

$$= A_3[1 + am(t)] + n_i(t). \qquad (3.4.6)$$

The average power in the term containing the message is thus $\langle A_3^2 a^2 m^2(t) \rangle$, so the output SNR is

$$\text{SNR}_o = \frac{a^2 A_3^2 \langle m^2(t) \rangle}{E\{n^2(t)\}} \qquad (3.4.7)$$

Comparing Eqs. (3.4.3) and (3.4.7), we find that

$$\text{SNR}_0 = 2\text{SNR}_i. \qquad (3.4.8)$$

Therefore, for high carrier-to-noise ratios, the envelope detector provides a 3-dB gain in output SNR, just as coherent detection does. It is important to note, however, that this result ignores the power contained in the carrier term alone, which is not necessary in AMDSB-SC.

For comparisons to other modulation methods such as FM, it is often convenient to express SNR_i, and SNR_0 in terms of the input carrier-to-noise ratio. It is evident from Eq. (3.4.1) that the average power in the carrier term alone is $A_3^2/2$, so that the input carrier-to-noise ratio, denoted CNR_i, is given by

$$\text{CNR}_i = \frac{A_3^2}{2E\{n^2(t)\}}. \tag{3.4.9}$$

Thus in terms of CNR_i, Eqs. (3.4.3) and (3.4.7) become

$$\text{SNR}_i = a^2\langle m^2(t)\rangle\text{CNR}_i \tag{3.4.10}$$

and

$$\text{SNR}_o = 2a^2\langle m^2(t)\rangle\text{CNR}_i, \tag{3.4.11}$$

respectively. Of course, Eq. (3.4.8) is unaffected.

For the small carrier-to-noise ratio $\left(A_3^2 \ll E\{\rho^2(t)\}\right)$, the envelope in Eq. (3.4.4) can be approximated as

$$\text{env}[r(t)] \cong \rho(t) + A_3[1 + \text{am}(t)]\cos\theta(t). \tag{3.4.12}$$

The message signal $m(t)$ is multiplied by the random variable $\cos\theta(t)$ and has added to it a very large amplitude random variable $\rho(t)$. The message is thus completely lost in the case of low carrier-to-noise ratio. There is some threshold value of CNR between the large and small CNR cases above which performance is acceptable and below which performance degrades very quickly. Such a threshold effect is common to nonlinear demodulation methods but is not evident for coherent demodulation of AMDSB-SC.

3.5 Frequency and Phase Modulation

To evaluate the performance of an FM system in the presence of additive narrowband Gaussian noise, we consider a standard FM waveform given by

$$s_{\text{FM}}(t) = A_c \cos\left[\omega_c t + c_f \int m(t)dt\right], \tag{3.5.1}$$

which is additively contaminated by the narrowband noise process $n(t)$ in Eq. (3.2.7) to produce the received signal

$$\begin{aligned}
r(t) &= s_{\text{FM}}(t) + n(t) \\
&= A_c \cos\left[\omega_c t + c_f \int m(t)dt\right] + \rho(t)\cos[\omega_c t + \theta(t)]. \tag{3.5.2}
\end{aligned}$$

The signal $r(t)$ is the waveform present at the IF filter output and at the input to the FM discriminator shown in Fig. 2.4.3. It will be useful to write Eq. (3.5.2) in the form

$$r(t) = A(t)\cos[\omega_c t + \phi(t)], \tag{3.5.3}$$

where $A(t)$ is of no interest due to the presence of the hard limiter in the discriminator. To gain some qualitative insight into the behavior of the phase $\phi(t)$, we let $\alpha(t) = \omega_c t + c_f \int m(t)dt$ (only for notational ease) and expand $r(t)$ in Eq. (3.5.2) as

$$r(t) = A_c \cos\alpha(t) + \rho(t) \cos\left[\alpha(t) + \theta(t) - c_f \int m(t)dt\right]$$

$$= A_c \cos\alpha(t) + \rho(t) \cos\left[\theta(t) - c_f \int m(t)dt\right] \cos\alpha(t)$$

$$- \rho(t) \sin\left[\theta(t) - c_f \int m(t)dt\right] \sin\alpha(t) \qquad (3.5.4)$$

Now, the phase with respect to the in-phase and quadrature components of $\alpha(t)$ can be found in the usual way from Eq. (3.5.4), so that $\phi(t)$ in Eq. (3.5.3) becomes

$$\phi(t) = c_f \int m(t)dt + \tan^{-1}\left\{\frac{\rho(t) \sin[\theta(t) - c_f \int m(t)dt]}{A_c + \rho(t) \cos[\theta(t) - c_f \int m(t)dt]}\right\}. \qquad (3.5.5)$$

For those readers comfortable with phasor notation, Eq. (3.5.5) is tantamount to using the instantaneous phase $c_f \int m(t)dt$ as a reference with the second term in Eq. (3.5.5) varying about this reference. Considering the high carrier-to-noise ratio case, where $A_c \gg \rho(t)$ we see from Eq. (3.5.5) that

$$\phi(t) \cong c_f \int m(t)dt + \frac{\rho(t)}{A_c} \sin\left[\theta(t) - c_f \int m(t)dt\right], \qquad (3.5.6)$$

where we have used the small-angle approximation on the arctan. Thus it is evident from Eq. (3.5.6) that in the large CNR case, the noise creates only a small random variation about the desired phase $c_f \int m(t)dt$ and is hence not much of a problem.

To examine the small CNR $[A_c \ll \rho(t)]$ behavior of $\theta(t)$, we follow an argument analogous to that employed to obtain Eq. (3.5.5) except that we use $\theta(t)$ as a reference [let $\alpha(t) = \omega_c t + \theta(t)$ and proceed] to find

$$\phi(t) = \theta(t) + \tan^{-1}\left\{\frac{A_c \sin[c_f \int m(t)dt - \theta(t)]}{\rho(t) + A_c \cos[c_f \int m(t)dt - \theta(t)]}\right\}. \qquad (3.5.7)$$

Invoking the small CNR assumption, we have that

$$\phi(t) \cong \theta(t) + \frac{A_c}{\rho(t)} \sin\left[c_f \int m(t)dt - \theta(t)\right]. \qquad (3.5.8)$$

Equation (3.5.8) reveals that for $A_c \ll \rho(t)$, the high-noise or weak-signal case, the phase is primarily the phase of the narrowband noise and the message signal cannot be recovered. The noise is sometimes said to *capture* the receiver in this case.

Besides obtaining an intuitive feel for the behavior of an FM receiver, the preceding analysis also makes it clear that our subsequent calculations of output SNR need only focus on the high-CNR case, because in the low-CNR situation, the system will be unusable, and hence of little practical interest.

To determine the output signal-to-noise ratio, we see from Fig. 2.4.3 that we need to differentiate the received signal and then pass it through an envelope detector. For purposes of analysis, this is equivalent to differentiating $\phi(t)$ in Eq. (3.5.6) and then low-pass filtering the result to the message signal bandwidth. Before proceeding, however, we need to simplify Eq. (3.5.6) still further. The term that causes the difficulty is the message component in the argument of the sine function. We know from Sect. 3.2 that $\phi(t)$ is uniformly distributed over the interval $[0, 2\pi]$, and thus if the message term were not present in the argument of the sine function, we could continue the analysis with little difficulty. Although we will not do so here, it can be shown that the presence of this message term serves only to produce noise outside the message bandwidth and hence can be neglected for the analysis here.

Therefore, returning to Eq. (3.5.6) and ignoring the message component in the sine function argument, we can approximate the additive phase of the carrier (for this analysis) by

$$
\phi(t) \cong c_f \int m(t)dt + \frac{\rho(t)}{A_c} \sin \theta(t)
$$

$$
= c_f \int m(t)dt + \frac{n_q(t)}{A_c}, \tag{3.5.9}
$$

where $n_q(t)$ is the white, zero-mean, Gaussian-distributed, quadrature component of the narrowband noise. Differentiating this $\phi(t)$ expression yields

$$
\frac{d}{dt}\phi(t) = c_f m(t) + \frac{1}{A_c}\frac{d}{dt}n_q(t). \tag{3.5.10}
$$

To find the mean-squared value of the second term in Eq. (3.5.10), we use results from random processes. The power spectral density of $n_q(t)$ is given by

$$
S_{n_q}(\omega) = \begin{cases} S_n(\omega - \omega_c) + S_n(\omega + \omega_c), & |\omega| \leq (\Delta\omega + \omega_m) \\ 0, & |\omega| > (\Delta\omega + \omega_m) \end{cases}, \tag{3.5.11}
$$

where ω_m is the bandwidth of the low-pass message $m(t)$, $\Delta\omega$ is the maximum frequency deviation, and $S_n(\omega)$ is the spectral density of the narrowband Gaussian process $n(t)$. If

$$
S_n(\omega) = \begin{cases} \mathcal{N}_0/2, & \omega_c - (\Delta\omega + \omega_m) \leq |\omega| \leq \omega_c + (\Delta\omega + \omega_m) \\ 0, & \text{otherwise}, \end{cases} \tag{3.5.12}
$$

then

$$S_{n_q}(\omega) = \mathcal{N}_0 \tag{3.5.13}$$

for $|\omega| \leq (\Delta\omega + \omega_m)$.

Passing this through the filter $H(\omega) = j\omega/A_c$, we get

$$S_{n_o}(\omega) = \frac{\omega^2}{A_c^2}\mathcal{N}_0, \quad |\omega| \leq \omega_m, \tag{3.5.14}$$

where we have defined

$$n_o(t) = \frac{1}{A_c}\frac{d}{dt}n_q(t)$$

The output noise power can finally be calculated as

$$E\{n_o^2(t)\} = \frac{1}{2\pi}\int_{-\omega_m}^{\omega_m}\frac{\omega^2\mathcal{N}_0}{A_c^2}d\omega = \frac{\omega_m^3\mathcal{N}_0}{3\pi A_c^2}. \tag{3.5.15}$$

Clearly, from Eq. (3.5.10), the output message or signal power is $c_f^2\langle m^2(t)\rangle$ so that the output signal-to-noise ratio for FM in the high-CNR case is

$$\text{SNR}_o = \frac{3\pi A_c^2 c_f^2\langle m^2(t)\rangle}{\omega_m^3\mathcal{N}_0}. \tag{3.5.16}$$

To determine the detection gain for FM, we need the input signal-to-noise ratio, which is the ratio of signal power to noise power for $r(t)$ in Eq. (3.5.2). Since for FM all of the transmitted power can be in the sidebands, we see that the signal power is from the first term in Eq. (3.5.2) given by $A_c^2/2$. Further, using Eq. (3.5.12), we obtain

$$E\{n^2(t)\} = \frac{\mathcal{N}_0(\Delta\omega + \omega_m)}{\pi}, \tag{3.5.17}$$

so that the input SNR is

$$\text{SNR}_i = \frac{\pi A_c^2}{2(\Delta\omega + \omega_m)\mathcal{N}_0}. \tag{3.5.18}$$

It is common in angle-modulated systems to call SNR_i the CNR. Thus, in terms of the CNR, we can rewrite Eq. (3.5.16) as

$$\text{SNR}_o = \frac{6c_f^2(\Delta\omega + \omega_m)\langle m^2(t)\rangle}{\omega_m^3}(\text{CNR}) \tag{3.5.19}$$

where the CNR = SNR_i in Eq. (3.5.18).

To draw specific conclusions concerning the detection gain, we need to consider a particular message signal $m(t)$ This is done in the following example.

Example 3.5.1 We wish to determine the output signal-to-noise ratio in terms of the CNR for the message signal

$$m(t) = A_m \cos \omega_m t.$$

In this case, Eq. (3.5.19) becomes

$$\mathrm{SNR}_o = \frac{3(\Delta\omega + \omega_m)c_f^2 A_m^2}{\omega_m^3}\mathrm{CNR}$$
$$= 3\beta^2(1+\beta)\mathrm{CNR}, \tag{3.5.20}$$

where we have used the facts that $\langle m^2(t) \rangle = A_m^2/2$ and that the modulation index $\beta = c_f A_m/\omega_m = \Delta\omega/\omega_m$. If we consider the important special case where $\beta = 5$, then

$$\mathrm{SNR}_o = 450\,\mathrm{CNR}. \tag{3.5.21}$$

Thus the detection gain for FM in the high-CNR case is 450 for a sinusoidal message signal and $\beta = 5$.

Since we know that the bandwidth of an FM signal for sinusoidal modulation is given by $\mathrm{BW} = 2\omega_m(1+\beta)$, we see that a linear increase in bandwidth results in an exponential improvement in detection gain (for a fixed value of CNR). This last parenthetical remark is extremely important, since it prevents us from concluding that we can increase SNR_O indefinitely by expanding bandwidth. This point is clarified by the following discussion. As the bandwidth is increased, more noise power appears at the detector input. Thus, to keep the CNR constant, the carrier power must be increased. Further, all practical communication systems have a constraint on the maximum available carrier power. Therefore, as we increase β to improve the detection gain, we expand bandwidth and thus admit more noise power into the receiver. To obtain the promised $3\beta^2(1 + \beta)$ improvement in detection gain, we must increase the carrier power in proportion to the noise power. However, we will usually operate at the maximum possible transmitted carrier power, and as a result, our detection gain is limited by this fixed carrier power constraint.

In FM systems, the message signal $m(t)$ is often preemphasized prior to modulation at the transmitter and then the demodulated signal is deemphasized at the receiver. Preemphasis consists of providing a gain for the components of $m(t)$ at the higher frequencies, while deemphasis attenuates the higher frequencies of the detected signal in an inverse proportion to the applied preemphasis. The utility of preemphasis/deemphasis can be made evident as follows. Most message signals, such as voice and music, have decreasing frequency content with increasing frequency. However, from Eq. (3.5.14), it is clear that the detector output noise power spectral density is increasing parabolically with increasing frequency. Therefore, the ratio of message signal power to noise power as a function of frequency decreases as frequency increases. The use of preemphasis boosts the higher frequencies in the message signal to try to compensate for the increased noise power. Of course, the preemphasis distorts the transmitted message signal, and therefore the inverse operation, deemphasis, must be performed on the demodulated message signal at the receiver. The deemphasis also reduces the output noise power at higher frequencies, which yields a better output SNR.

Virtually all of our analyses in this section have been for the high-CNR case, since for the small-CNR case, the demodulated output is due primarily to the noise alone. As the CNR is decreased from a large value to smaller values, there is a point at which there is a precipitous drop in SNR_o if the CNR is decreased still further. This value of CNR is called the *threshold,* and although it varies as a function of modulation index β the threshold value of CNR is roughly 10 dB. Figure 3.3 shows a plot of SNR_o versus CNR as a function of β for a typical FM system. The threshold effect is clearly evident. None

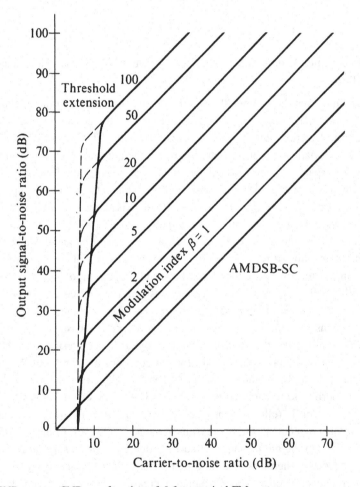

Fig. 3.3 SNR_o versus CNR as a function of β for a typical FM system

of the analyses presented here indicate why the threshold changes with β. However, the effect is explained qualitatively as follows. For small values of β the FM signal tends to occupy the entire bandwidth $2\omega_m(1 + \beta)$. As β is increased, the bandwidth of the FM signal is also increased, but at any time instant, only a small portion of the FM bandwidth is occupied by the transmitted signal, which sweeps through the wider bandwidth in relation to the message signal. Therefore, at any time instant, the (fixed) signal power occupies a smaller percentage of the intermediate-frequency bandwidth of $2\omega_m(1 + \beta)$ as β is increased, thus requiring a higher CNR to stay above the threshold.

The threshold can be made independent of β, as indicated by the dashed lines in Fig. 3.3, by using what is called FM feedback (FMFB) or a phase-locked loop (PLL). In general, these systems cause the demodulator to track the received signal as a function of time, thus allowing a narrower IF bandwidth to be used. As a result, as β is increased, the IF bandwidth does not increase, and greater noise rejection is achieved.

For phase modulation, the analysis is quite similar to that for FM, but the results are significantly different. The transmitted waveform for PM is

$$s_{\text{PM}}(t) = A_c \cos\left[\omega_c t + c_p m(t)\right] \tag{3.5.22}$$

and the received signal, after contamination with narrowband, white Gaussian noise is

$$\begin{aligned} r(t) &= s_{\text{PM}}(t) + n(t) \\ &= A_c \cos\left[\omega_c t + c_p m(t)\right] + \rho(t)\cos[\omega_c t + \theta(t)]. \end{aligned} \tag{3.5.23}$$

As before, we can write Eq. (3.5.23) in the form

$$r(t) = A(t)\cos[\omega_c t + \phi(t)], \tag{3.5.24}$$

where for the large CNR case, $A_c \gg \rho(t)$,

$$\begin{aligned} \phi(t) &\cong c_p m(t) + \frac{\rho(t)}{A_c}\sin\left[\theta(t) - c_p m(t)\right] \\ &\cong c_p m(t) + \frac{\rho(t)}{A_c}\sin\theta(t) \\ &= c_p m(t) + \frac{n_q(t)}{A_c}. \end{aligned} \tag{3.5.25}$$

It is now straightforward to show that the output signal-to-noise ratio for PM is

$$\text{SNR}_o = \frac{\pi A_c^2 c_p^2 \langle m^2(t)\rangle}{\omega_m N_0} \tag{3.5.26}$$

The input SNR or CNR is still given by Eq. (3.5.18), so in terms of the CNR we have

$$\text{SNR}_o = \frac{2(\Delta\omega + \omega_m)c_p^2\langle m^2(t)\rangle}{\omega_m}(\text{CNR}) \tag{3.5.27}$$

for PM.

3.6   SNR Comparisons for Systems

We now attempt to pool the results of the preceding sections to obtain a relative comparison of the several analog communication systems discussed in this chapter. We use the output signal-to-noise ratio as our performance indicator, and for our comparisons to be meaningful, we require that all the systems have the same ratio of *total* transmitted power to received noise power. We denote the latter ratio by SNR_R. Thus the comparisons here are not based on detection gain but on SNR_o for a fixed SNR_R. Of course, bandwidth is another important parameter when comparing communication systems, and we state the bandwidth as some multiple of the message waveform bandwidth ω_m. Finally, when necessary, we assume a particular message signal, namely single-tone modulation given by $m(t) = A_m \cos \omega_m t$. There is some loss of generality in this last assumption, but in order to calculate $\langle m^2(t)\rangle$, we must have a specific $m(t)$ and a pure tone is a common reference signal.

In Sects. 3.3–3.5, we found the output signal-to-noise ratio as a function of input SNR, where SNR_i is the ratio of signal power in the *message* component to the noise power. Note the difference between SNR_i and SNR_R, whose numerator is the *total* transmitted power. Thus we shall find the SNR_i in terms of SNR_R for each of the systems of interest, and then substitute the SNR_i value into the appropriate expressions in the preceding sections to obtain SNR_o in terms of SNR_R.

Beginning with AMDSB-SC, the transmitted signal is

$$s_{\text{DSB}}(t) = A_1 m(t) \cos \omega_c t, \tag{3.6.1}$$

where we do not as yet use the assumption that $m(t) = A_m \cos \omega_m t$. For noise with a flat power spectral density of amplitude $\mathcal{N}_0/2$ in $\omega_c - \omega_m \leq |\omega| \leq \omega_c + \omega_m$,

$$\text{SNR}_{R,\text{DSB}} = \frac{\left(A_1^2/2\right)\langle m^2(t)\rangle}{(\mathcal{N}_0/2)(4\omega_m)/2\pi} = \frac{2\pi A_1^2\langle m^2(t)\rangle}{4\mathcal{N}_0\omega_m}. \tag{3.6.2}$$

Since, for DSB-SC, all of the transmitted power is in the sidebands, it is evident that $\text{SNR}_{i,\text{DSB}} = \text{SNR}_{R,\text{DSB}}$ [see Eq. (3.3.3)], so from Eq. (3.3.6),

$$\text{SNR}_{o,\text{DSB}} = 2\text{SNR}_{R,\text{DSB}} = 2\text{SNR}_R, \tag{3.6.3}$$

where we have defined $\text{SNR}_R \triangleq \text{SNR}_{R,DSB}$ as our fixed reference.

For AMSSB-SC, the transmitted signal can be written as

$$s_{\text{SSB}}(t) = A_2 m(t) \cos \omega_c t \pm A_2 m_h(t) \sin \omega_c t, \tag{3.6.4}$$

so that the total transmitted power is $A_2^2 \langle m^2(t) \rangle$ [see Eq. (3.3.8)]. We choose A_2 such that the total transmitted power for SSB is the same as for DSB, namely $A_1^2 \langle m^2(t) \rangle / 2$ Only half the bandwidth of DSB is required, so that the received noise power is $(1/2\pi)(N_0/2)(2\omega_m) = N_0 \omega_m / 2\pi$, and therefore,

$$\text{SNR}_{R,\text{SSB}} = \frac{2\pi A_2^2 \langle m^2(t) \rangle}{N_0 \omega_m} = \frac{2\pi A_1^2 \langle m^2(t) \rangle}{2N_0 \omega_m}$$

$$= 2\text{SNR}_R. \tag{3.6.5}$$

Since for SSB all of the transmitted power is in the sidebands, $\text{SNR}_{i,\text{SSB}} = \text{SNR}_{R,\text{SSB}}$, so using Eq. (3.3.12), we have that

$$\text{SNR}_{o,\text{SSB}} = 2\text{SNR}_R. \tag{3.6.6}$$

The transmitted waveform for AMDSB-TC is

$$s_{\text{AM}}(t) = A_3[1 + am(t)] \cos \omega_c t \tag{3.6.7}$$

[$m(t)$ normalized] with total power $A_3^2 [1 + a^2 \langle m^2(t) \rangle]/2$, where we have assumed that $\langle m(t) \rangle = 0$. Again, we must keep the total transmitted power the same in all cases, so we adjust A_3 to obtain a total power of $A_1^2 \langle m^2(t) \rangle / 2$. The received noise power is the same for the suppressed carrier case as for the transmitted carrier case, so

$$\text{SNR}_{\text{AM}} = \frac{A_3^2[1 + a^2 \langle m^2(t) \rangle]/2}{2N_0 \omega_m / 2\pi}$$

$$= \frac{2\pi A_1^2 \langle m^2(t) \rangle}{4N_0 \omega_m} = \text{SNR}_R. \tag{3.6.8}$$

Now, from Eq. (3.4.3),

$$\text{SNR}_{i,\text{AM}} = \frac{2\pi a^2 A_3^2 \langle m^2(t) \rangle}{4N_0 \omega_m}$$

$$= \frac{a^2 \langle m^2(t) \rangle}{1 + a^2 \langle m^2(t) \rangle} \text{SNR}_R, \tag{3.6.9}$$

so that using Eq. (3.4.8), we get

$$\text{SNR}_{o,\text{AM}} = \frac{2a^2 \langle m^2(t) \rangle}{1 + a^2 \langle m^2(t) \rangle} \text{SNR}_R. \tag{3.6.10}$$

Using Eqs. (3.6.3), (3.6.6), and (3.6.10), we are now in a position to compare the performance of these three AM systems. We see that AMDSB-SC and AMSSB-SC perform equally well, but since $0 < a \leq 1$ and we have assumed that $m(t)$ is normalized here, $a^2 \langle m^2(t) \rangle / [1 + a^2 \langle m^2(t) \rangle] < 1$, the performance of AMDSB-TC is poorer than the other two systems. Of course, this is an expected result, since in AMDSB-TC we allocate power to a separate carrier term to simplify the receiver. If we ignore complexity, AMSSB-SC would be the best choice among these three systems, due to its least bandwidth requirements.

To compare FM and PM with the AM systems, we continue to use the assumption that $m(t)$ is normalized and consider Eqs. (3.5.19) and (3.5.27) for FM and PM, respectively. For FM we substitute Eq. (3.5.18) into (3.5.19), which yields

$$\text{SNR}_{o,\text{FM}} = \frac{3\pi A_c^2 c_f^2 \langle m^2(t) \rangle}{\mathcal{N}_0 \omega_m^3}$$

$$= \frac{3\beta^2 (2\pi A_c^2) \langle m^2(t) \rangle}{2 \mathcal{N}_0 \omega_m}, \tag{3.6.11}$$

where since $|m(t)| \leq 1$, we have let $\Delta\omega = c_f$. With reference to Eq. (3.6.2), we let $A_c = A_1$, so from Eq. (3.6.3) we see that

$$\text{SNR}_{o,\text{FM}} = 3\beta^2 \text{SNR}_{o,\text{DSB}}, \tag{3.6.12}$$

and it follows from Eq. (3.6.6) that

$$\text{SNR}_{o,\text{FM}} = 3\beta^2 \text{SNR}_{o,\text{SSB}}. \tag{3.6.13}$$

Further, it follows directly from Eq. (3.6.10) that

$$\text{SNR}_{o,\text{FM}} = 3\beta^2 \frac{[1 + a^2 \langle m^2(t) \rangle]}{a^2 \langle m^2(t) \rangle} \text{SNR}_{o,\text{AM}}. \tag{3.6.14}$$

It is often convenient to compare FM and conventional AM in terms of the carrier-to-noise ratio. To do this, we can use Eq. (3.4.9) with $A_3 = A_c$ or Eq. (3.6.9) with $A_3 = A_c$ to produce

$$\text{SNR}_{o,\text{AM}} = a^2 \langle m^2(t) \rangle \text{CNR}_{\text{AM}}, \tag{3.6.15}$$

where $\text{CNR}_{\text{AM}} \triangleq 2\pi A_c^2 / 2\mathcal{N}_0 \omega_m$. Thus

$$\text{SNR}_{o,\text{FM}} = 3\beta^2 \langle m^2(t) \rangle \text{CNR}_{\text{AM}}. \tag{3.6.16}$$

Under the same assumptions used for FM, it is straightforward to show that for PM,

$$\text{SNR}_{o,\text{PM}} = c_p^2 \text{SNR}_{o,\text{DSB}} = c_p^2 \text{SNR}_{o,\text{SSB}} \tag{3.6.17}$$

and

$$\text{SNR}_{o,\text{PM}} = c_p^2 \frac{\left[1 + a^2 \langle m^2(t) \rangle\right]}{a^2 \langle m^2(t) \rangle} \text{SNR}_{o,\text{AM}}. \tag{3.6.18}$$

It is evident from Eqs. (3.6.12)–(3.6.18) that for $\beta \gg 1$ (wideband FM) that the improvement in SNR provided by FM over AM can be substantial. Of course, the larger β is, the larger the FM or PM bandwidth requirements.

It is difficult to draw general conclusions concerning the relative performance of FM and PM, since their performance varies depending on the characteristics of the message signal. We give one comparison of interest in the following example.

Example 3.6.1 In this example we compare the performance of AMDSB-SC, AMDSB-TC, AMSSB-SC, FM, and PM for the particular case of a sinusoidal message signal $m(t) = \cos \omega_m t$. Clearly, Eqs. (3.6.12), (3.6.13), (3.6.17) still hold and are unaffected by the choice of modulating signal. However, for conventional AM we use the fact that $\langle m^2(t) \rangle = \frac{1}{2}$ and assume that $a = 1$, to obtain

$$\text{SNR}_{o,\text{FM}} = 9\beta^2 \text{SNR}_{o,\text{AM}}. \tag{3.6.19}$$

Similarly for PM,

$$\text{SNR}_{o,\text{PM}} = 3c_p^2 \text{SNR}_{o,\text{AM}}. \tag{3.6.20}$$

To compare FM and PM, we note that here $\beta = c_f / \omega_m$, so that

$$\text{SNR}_{o,\text{FM}} = \frac{3\beta^2}{c_p^2} \text{SNR}_{o,\text{PM}} = \frac{3c_f^2}{\omega_m^2 c_p^2} \text{SNR}_{o,\text{PM}}. \tag{3.6.21}$$

Thus, if $c_f^2 / \omega_m^2 c_p^2 > \frac{1}{3}$, $\text{SNR}_{o,\text{FM}} > \text{SNR}_{o,\text{PM}}$. This is often true for nonbaseband (bandpass) signals.

Summary
In this chapter we have analyzed the performance of analog modulation systems with analog messages first by finding the improvement in signal-to-noise ratio afforded by the demodulation process and then by comparing systems for the same ratio of total transmitted power to noise power. The advantage of coherent detection is clear, and the threshold effects in noncoherent detection of conventional AM and in FM/PM are evident. The performance improvement provided by FM over AM has been demonstrated. Considerably more detailed analyses are possible, but the SNR comparisons in Sect. 3.6 coupled with a knowledge of complexity and bandwidth requirements for the various methods allow competent design trade-offs to be made.

Problems

3.1   Derive the joint pdf of $\rho(t)$ and $\theta(t)$ in Eq. (3.2.12).

3.2   Derive expressions for $n_i(t)$ and $n_q(t)$ in terms of $n(t)$ and its Hilbert transform $n_h(t)$.

3.3   Show that if $n(t)$ is a WSS random process, $n(t)$ and its Hilbert transform have the same autocorrelation function.

3.4   Given the WSS narrowband noise $n(t)$ and its Hilbert transform $n_h(t)$, show that:
      (a) $R_{nn_h}(\tau) = \mathcal{H}[R_n(\tau)]$.
      (b) $R_{n_h n}(\tau) = -\mathcal{H}[R_n(\tau)]$.

3.5   Show that the in-phase and quadrature components of WSS narrowband noise have the same autocorrelation function. This is Property 3.
      Hint: See previous problems.

3.6   Derive the cross-correlation function between the in-phase and quadrature components of narrowband noise.

3.7   Derive the spectral densities of the in-phase and quadrature components of WSS narrowband noise.

3.8   Given the power spectral density of narrowband noise $S_n(\omega)$ in Fig. 3.4, sketch $S_{n_i}(\omega)$ and $S_{n_q}(\omega)$

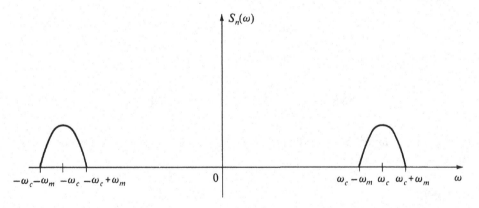

Fig. 3.4 Power Spectral Density of Narrowband Noise for Problem 3.8

3.9   Wide-sense stationary white noise with two-sided power spectral density of amplitude $\mathcal{N}_0/2$ is applied to an ideal bandpass filter $H(\omega)$ with $|H(\omega)| = 1$ for $\omega_c - \omega_B \leq |\omega| \leq \omega_c + \omega_B$ and $\angle H(\omega) = 0$ Find the autocorrelation functions of the filter output in-phase and quadrature components.

3.10 Validate Eq. (3.3.8).

3.11   Derive Eq. (3.4.12).

3.12   Validate Eqs. (3.5.6) and (3.5.8).

3.13   Sketch the noise spectral density in Eq. (3.5.14). What does this imply?

3.14   Plot SNR_o in Eq. (3.5.20) versus required bandwidth.

3.15   Plot Eq. (3.6.10) versus a if $\langle m^2(t) \rangle = \frac{1}{2}$.

3.16   Plot SNR_o versus SNR_R for Eqs. (3.6.3), (3.6.6), and (3.6.10) on the same sheet. Let $a = 1$ and $\langle m^2(t) \rangle = \frac{1}{2}$. Develop an expression relating SNR_o and SNR_R for FM and plot it on the same sheet. Let $\beta = 5$.

3.17   For what range of β does FM offer an SNR_o improvement over AMDSB-SC? Over conventional AM if $a = 1$ and $\langle m^2(t) \rangle = \frac{1}{2}$?

3.18   A conventional AM signal can be demodulated by a squarer followed by a low-pass filter. If the input to the square-law detector is a conventional AM signal plus narrowband noise, find an expression for SNR_o in terms of SNR_i. Compare your result to the result for envelope detection in Eq. (3.6.10) by letting $a = 1$ and $\langle m^2(t) \rangle = \frac{1}{2}$.

3.19   Use Eq. (3.6.21) and compare FM and PM systems for the same β. Repeat the comparison if both have the same $\Delta\omega$. Which comparison seems more practical?

3.20   It is mentioned in Sect. 3.5 that preemphasis/deemphasis is often used to advantage in FM systems. If the preemphasis network has the transfer function $H(\omega) = 1 + j\omega/\omega_0$, find an expression for the output noise power in FM using preemphasis/deemphasis.

3.21   Use the result of Problem 3.20 to plot the reduction in noise power as a function of the ratio ω_m/ω_0, where ω_m is the message signal bandwidth. For commercial FM $\omega_m = 2\pi f_m = 2\pi(15\text{kHz})$ and $\omega_0 = 2\pi f_0 = 2\pi$ (2.1 kHz). What is the reduction in noise power for commercial FM using preemphasis/deemphasis?

3.22   Show that preemphasis/deemphasis can also be used to provide a performance improvement for conventional AM.

3.23   Refer to Fig. 1.3.3 of an envelope detector and the corresponding discussion. Use this development to explain heuristically the threshold effect in an envelope detector for conventional AM in the small carrier-to-noise ratio case.

3.24   The specific definition of the threshold between the large and small carrier- to-noise ratio cases is difficult and has been avoided thus far. Carlson (1975) suggests that it is that value of SNR_i such that the probability of the noise envelope exceeding the carrier amplitude is 0.01. With reference to Eq. (3.4.12), this is $P[\rho(t) > A_3] = 0.01$. Find that SNR_i if $a = 1$ and $\langle m^2(t) \rangle = \frac{1}{2}$.

3.25   Discuss why the output signal-to-noise ratio in stereo FM is worse than in monophonic FM.

3.26   The threshold phenomenon in FM occurs when $A_c \cong \rho(t)$ in Eqs. (3.5.2), (3.5.4), and (3.5.5). In this situation, if we think of A_c and $\rho(t)$ as phasors, $\rho(t)$ can cause a complete 2π phase change in $\phi(t)$ in Eqs. (3.5.3) and (3.5.4), which yields an impulse in $d\phi(t)/dt$. These impulses, called "clicks," have been analyzed by Rice (1963) and are very detrimental to FM system performance (Taub and Schilling, 1971), The output signal-to-noise ratio, including the effects of clicks for a sinusoidal message signal, is

$$\text{SNR}_o = \frac{\frac{3}{2}\beta^2 \text{SNR}_i}{1 + (12\beta/\pi)\text{SNR}_i e^{-\text{SNR}_i/(1+\beta)2}}.$$

Plot SNR_o versus SNR_i for $\beta = 5, 10,$ and 15.

3.27. Use the results of Problem 3.26 to show that:

  (a) For a fixed SNR_i, exchanging bandwidth for SNR_o is not always possible.

  (b) For a fixed SNR_o, exchanging bandwidth for a reduced SNR_i is not always possible.

3.28. Assume that the narrowband noise has the power spectral density shown in Fig. 3.5. Find $S_{n_i}(\omega)$, $S_{n_q}(\omega)$, $R_{n_q}(\tau)$, $R_{n_i}(\tau)$, $E\left[n_i^2(t)\right]$ and $E\left[n_q^2(t)\right]$.

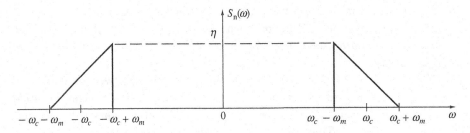

Fig. 3.5 Narrowband Noise Power Spectral Density for Problem 3.28

3.29. Compare $\text{SNR}_{o,\text{DSB}}$ and $\text{SNR}_{o,\text{SSB}}$ for narrowband noise with the power spectral density shown in Fig. 3.5.

3.30. Does a nonflat noise power spectral density as in Fig. 3.5 change the FM SNR_o analysis leading to Eq. (3.5.19)? If so, how?

Bibliography

Abramson, N. 1973. "The Aloha System." Chapter 14 in *Computer Communication Networks*. N. Abramson and F. F. Kuo, eds. Englewood Cliffs, N.J.: Prentice Hall, pp. 501–517.

Amoroso, F. 1980. "The Bandwidth of Digital Data Signals." *IEEE Commun. Mag.*, Vol. 18, Nov.

Anderson, J. B., and J. R. Lesh, eds. 1981. "Special Section on Combined Coding and Modulation." *IEEE Trans. Commun.*, Vol. COM-29, Mar.

Anderson, J. B, C.-E. W. Sundberg, T. Aulin, and N. Rydbeck. 1981. "Power-Bandwidth Performance of Smoothed Phase Modulation Codes," *IEEE Trans. Commun.*, Vol. COM-29, Mar., pp. 187–195.

Anderson, R. R., and J. Salz. 1965. "Spectra of Digital FM." Bell Syst. Tech. J., Vol. 44, July–Aug., pp. 1165–1189.

AT&T, Telecommunications Transmission Engineering, Vols. 1–3. New York: AT&T, 1977.

Bellamy, J. 1982. *Digital Telephony*. New York: Wiley.

Berger, T. 1971. Rate Distortion Theory: A Mathematical Basis for Data Compression. *Englewood Cliffs, N.J.: Prentice Hall.*

Bertsekas, D., and R. Gallager. 1987. Data Networks. Englewood Cliffs, N.J.: Prentice Hall.

Bhargava, V. K., D. Haccoun, R. Matyas, and P. Nuspl. 1981. *Digital Communications by Satellite.* New York: Wiley.

Blahut, R. E. 1983. *Theory and Practice of Error Control Codes*. Reading, Mass.: Addison-Wesley.

Blahut, R. E. 1990. *Digital Transmission of Information*. Reading, Mass.: Addison-Wesley.

Briley, B. E. 1983. *Introduction to Telephone Switching*. Reading, Mass.: Addison-Wesley.

Capon, J. 1959. "A Probabilistic Model for Run-Length Coding of Pictures." *IRE Trans. Inf. Theory,* Vol. IT-5, pp. 157–163.

Carlson, A. B. 1975. *Communication Systems*. New York: McGraw-Hill.

Chen, W. H. 1963. *The Analysis of Linear Systems*. New York: McGraw-Hill.

Clark, G. C., Jr., and J. B. Cain. 1981. *Error-Correction Coding for Digital Communications*. New York: Plenum Press.

Clos, C. 1953. "A Study of Non-blocking Switching Networks," *Bell Syst. Tech. J.,* Vol. 32, Mar., pp. 406–424.

Collins, A. A., and R. D. Pedersen. 1973. *Telecommunications: A Time for Innovation.* Dallas, Tex.: Merle Collins Foundation.

Cooper, G. R., and C. D. McGillem. 1986. *Modem Communications and Spread Spectrum*. New York: McGraw-Hill.

Davies, D. W., and D. L. A. Barber. 1973. *Communication Networks for Computers*. New York: Wiley.

© The Editor(s) (if applicable) and The Author(s), under exclusive license to Springer Nature Switzerland AG 2023
J. D. Gibson, *Analog Communications*, Synthesis Lectures on Communications, https://doi.org/10.1007/978-3-031-19584-6

de Buda, R. 1972. "Coherent Demodulation of Frequency Shift Keying with Low Deviation Ratio." *IEEE Trans. Commun.,* Vol. COM-20, June, pp. 429–436.

Dixon, R. C. 1976. *Spread Spectrum Techniques.* New York: IEEE Press.

Dixon, R. C. 1984. *Spread Spectrum Systems.* New York: Wiley.

Farvardin, N., and J. W. Modestino. 1984. "Optimum Quantizer Performance for a Class of Non-Gaussian Memoryless Sources." *IEEE Trans. Inf. Theory,* Vol. IT-30, May, pp. 485–497.

Forney, G. David, Jr. 1989. "Introduction to Modem Technology: Theory and Practice of Bandwidth Efficient Modulation from Shannon and Nyquist to Date." University Video Communications and the IEEE; Copyright: Motorola.

Foschini, G. J., R. D. Gitlin, and S. B. Weinstein. 1974. "Optimization of Two-Dimensional Signal Constellations in the Presence of Gaussian Noise." *IEEE Trans. Commim.,* Vol. COM-22, Jan., pp. 28–38.

Franks, L. E. 1969. *Signal Theory.* Englewood Cliffs, N. J.: Prentice Hall.

Franks, L. E. 1980. "Carrier and Bit Synchronization in Data Communication: A Tutorial Review." *IEEE Trans. Commun.,* Vol. COM-28, Aug., pp. 1107–1121.

Freeman, R. L. 1981. *Telecommunications Transmission Handbook,* 2nd ed. New York: Wiley.

Gallager, R. G. 1968. Information Theory and Reliable Communication. New York: Wiley.

Gersho, A., and R. M. Gray. 1991. *Vector Quantization and Signal Compression.* Hingham, Mass.: Kluwer.

Gilhousen, K. S., I. M. Jacobs, R. Padovani, and L. A. Weaver, Jr. 1990. "Increased Capacity Using CDMA for Mobile Satellite Communication." *IEEE J. Sel. Areas Commun.,* Vol. SAC-8, May, pp. 503–514.

Gold, R. 1967. "Optimal Binary Sequences for Spread-Spectrum Multiplexing." *IEEE Trans. Inf. Theory,* Vol. IT-13, pp. 619–621.

Gold, R. 1968. "Maximal Recursive Sequences with 3-Valued Recursive Cross Correlation Functions." *IEEE Trans. Inf. Theory,* Vol. IT-14, Jan., pp. 154–156.

Golomb, S. W., ed. 1964. *Digital Communications with Space Applications.* Englewood Cliffs, N.J.: Prentice Hall.

Gray, R. M., and L. D. Davisson. 1986. *Random Processes: A Mathematical Approach for Engineers.* Englewood Cliffs, N.J.: Prentice Hall.

Hammond, J. L., and P. J. P. O'Reilly. 1986. *Performance Analysis of Local Computer Networks.* Reading, Mass.: Addison-Wesley.

Haykin, S. 1983. *Communication Systems.* New York: Wiley.

Heller, J. A., and I. M. Jacobs. 1971. "Viterbi Decoding for Satellite and Space Communications." *IEEE Trans. Commun. Technol,* Vol. COM-19, Oct., pp. 835–848.

Hirsch, D., and W. J. Wolf. 1970. "A Simple Adaptive Equalizer for Efficient Data Transmission." *IEEE Trans. Commun. Technol.,* Vol. COM-18, Feb., pp. 5–12.

Holmes, J. K. 1982. *Coherent Spread Spectrum Systems.* New York: Wiley.

Holzman, L. N, and W. J. Lawless. 1970. "Data Set 203: A New High-Speed Voiceband Modem." *Computer,* Sept.-Oct., pp. 24–30.

Houston, S. W. 1975. "Modulation Techniques for Communication, Part I: Tone and Noise Jamming Performance for Spread Spectrum M-ary FSK and 2, 4-ary DPSK Waveforms." Proceedings of the IEEE National Aerospace and Electronics Conference (NAECON '75), Dayton, Ohio, June 10–12, pp. 51–58.

Huffman, D. A. 1952. "A Method for the Construction of Minimum Redundancy Codes." *Proc. IRE,* Vol. 40, Sept., pp. 1098–1101.

Jackson, D. 1941. *Fourier Series and Orthogonal Polynomials.* Washington, D.C.: The Mathematical Association of America.

Jacobaeus, C. 1950. "A Study of Congestion in Link Systems." *Ericsson Tech.,* No. 48, Stockholm.

Jahnke, E., and F. Emde. 1945. *Tables of Functions.* New York: Dover.

Jayant, N. S., and P. Noll. 1984. *Digital Coding of Waveforms.* Englewood Cliffs, N.J.: Prentice Hall.

Johnson, G. D. 1973. "No. 4 ESS." *Bell Lab. Rec.,* Sept., pp. 226–232.

Kaplan, W. 1959. *Advanced Calculus.* Reading, Mass.: Addison-Wesley.

Keiser, G. E. 1989. *Local Area Networks.* New York: McGraw-Hill.

Kernighan, B. W., and S. Lin. 1973. "Heuristic Solution of a Signal Design Optimization Problem." *Proc. 7th Annual Princeton Conference on Information Science and Systems,* Mar.

Kleinrock, L., and S. S. Lam. 1975. "Packet Switching in a Multiaccess Broadcast Channel: Performance Evaluation." *IEEE Trans. Commun.,* Vol. COM-23, Apr., pp. 410–423.

Kotel'nikov, V. A. 1947. *The Theory of Optimum Noise Immunity.* Doctoral dissertation, Molotov Energy Institute, Moscow. Also published by McGraw-Hill, New York, 1959.

Kretzmer, E. R. 1965. "Binary Data Communication by Partial Response Transmission." *Conf. Rec.,* 1965 IEEE Annual Communications Conference, pp. 451–455.

Kretzmer, E. R. 1966. "Generalization of a Technique for Binary Data Communication." *IEEE Trans. Commun. Technol.,* Feb., pp. 67–68.

Kuo, F. F. 1962. *Network Analysis and Synthesis.* New York: Wiley.

Lathi, B. P. 1968. *Communication Systems.* New York: Wiley.

Lee, C. Y. 1955. "Analysis of Switching Networks." *Bell Syst. Tech. J.,* Vol. 34, Nov., pp. 1287–1315.

Lender, A. 1963. "The Duobinary Technique for High Speed Data Transmission." *IEEE Trans. Commun. Electron.,* Vol. 82, May, pp. 214–218.

Lender, A. 1964. "Correlative Digital Communication Techniques." *IEEE Trans. Commun. Technol,* Dec., pp. 128–135.

Lender, A. 1966. "Correlative Level Coding for Binary Data Transmission." *IEEE Spectrum,* Vol. 3, Feb., pp. 104–115.

Lender, A. 1981. Chapter 7 in *Digital Communications: Microwave Applications.* K. Feher, ed. Englewood Cliffs, N.J.: Prentice Hall, pp. 144–182.

Levitt, B. K. 1985. "Strategies for FH/MFSK Signaling with Diversity in Worst-Case Partial Band Noise." *IEEE J. Sel. Areas Commun.,* Vol. SAC-3, Sept., pp. 622–626.

Lin, S., and D. J. Costello, Jr. 1983. *Error Control Coding: Fundamentals and Applications.* Englewood Cliffs, N.J.: Prentice Hall.

Lloyd, S. P. 1982. "Least Squares Quantization in PCM." *IEEE Trans. Inf. Theory,* Vol. IT-28, Mar., pp. 129–137 (unpublished memorandum, Bell Laboratories, 1957).

Lucky, R. W. 1965. "Automatic Equalization for Digital Communications." *Bell Syst. Tech. J.,* Vol. 44, Apr., pp. 547–588.

Lucky, R. W. 1966. "Techniques for Adaptive Equalization of Digital Communication." *Bell Syst. Tech. J.,* Vol. 45, Feb., pp. 255–286.

Lucky, R. W., and H. Rudin. 1967. "An Automatic Equalizer for General-Purpose Communication Channels." *Bell Syst. Tech. J.,* Vol. 46, Nov., pp. 2179–2207.

Lucky, R. W., J. Salz, and E. J. Weldon, Jr. 1968. *Principles of Data Communication.* New York: McGraw-Hill.

Makhoul, J. 1975. "Linear Prediction: A Tutorial Review." *Proc. IEEE,* Vol. 63, Apr., pp. 561–580.

Martin, D. R., and P. L. McAdam. 1980. "Convolutional Code Performance with Optimal Jamming." *Conf. Rec.,* 1980 IEEE International Conference on Communications, Seattle, Wash., June 8–12, pp. 4.3.1–4.3.7.

Max, J. 1960. "Quantizing for Minimum Distortion." *IRE Trans. Inf. Theory,* Vol. IT-6, Mar., pp. 7–12.

Mazur, B. A., and D. P. Taylor. 1981. "Demodulation and Carrier Synchronization of Multi-h Phase Codes." *IEEE Trans. Commun.,* Vol. COM-29, Mar., pp. 257–266.

McEliece, R. J. 1977. *The Theory of Information and Coding.* Reading, Mass.: Addison-Wesley.

Newman, D. B., Jr., and R. L. Pickholtz, eds. 1987. Special Issue on "Network Security." *IEEE Network,* Vol. 1, Apr.

Noll, P., and R. Zelinski. 1978. "Bounds on Quantizer Performance in the Low Bit-Rate Region." *IEEE Trans. Commun.,* Vol. COM-26, Feb., pp. 300–304.

Nyquist, H. 1924. "Certain Factors Affecting Telegraph Speed." *Bell Syst. Tech. J.,* Vol. 3, Apr., pp. 324–346.

Nyquist, H. 1928. "Certain Topics in Telegraph Transmission Theory." *Trans. AIEE,* Vol. 47, Apr, pp. 617–644.

Owen, F. F. E. 1982. *PCM and Digital Transmission Systems.* New York: McGraw-Hill.

Paez, M. D., and T. H. Glisson. 1972. "Minimum Mean-Squared-Error Quantization in Speech PCM and DPCM Systems." *IEEE Trans. Commun.,* Vol. COM-20, Apr., pp. 225–230.

Pahlavan, K, and J. L. Holsinger. 1988. "Voice-Band Data Communication Modems: A Historical Review: 1919–1988." *IEEE Commun. Mag.,* Vol. 26, Jan, pp. 16–27.

Pasupathy, S. 1977. "Correlative Coding: A Bandwidth-Efficient Signaling Scheme." *IEEE Commun. Mag.,* Vol. 15, July, pp. 4–11.

Pasupathy, S. 1979. "Minimum Shift Keying: A Spectrally Efficient Modulation." *IEEE Commun. Mag.,* Vol. 17, July, pp. 14–22.

Pickholtz, R. L, D. L. Schilling, and L. B. Milstein. 1982. "Theory of Spread-Spectrum Communications: A Tutorial." *IEEE Trans. Commun.,* Vol. COM-30, May, pp. 855–884.

Pickholtz, R. L, D. L. Schilling, and L. B. Milstein. 1984. "Revisions to 'Theory of Spread Spectrum Communications: A Tutorial'," *IEEE Trans. Commun.,* Vol. COM-32, Feb, pp. 211–212.

Pratt, W. K. 1978. *Digital Image Processing.* New York: Wiley.

Proakis, J. G. 1989. *Digital Communications.* New York: McGraw-Hill.

Qureshi, S. U. H. 1985. "Adaptive Equalization." *Proc. IEEE,* Vol. 73, Sept., pp. 1349–1387.

Rabiner, L. R., and R. W. Schafer. 1978. *Digital Processing of Speech Signals.* Englewood Cliffs, N.J.: Prentice Hall.

Rice, S. O. 1963. "Noise in FM Receivers." Chapter 25 in *Proc., Symposium on Time Series Analysis.* M. Rosenblatt, ed. New York: Wiley, pp. 395–424.

Rice, S. O. 1982. "Envelopes of Narrow-Band Signals." *Proc. IEEE,* Vol. 70, July, pp. 692–699.

Sarwate, D. V., and M. B. Pursley. 1980. "Crosscorrelation Properties of Pseudorandom and Related Sequences." *Proc. IEEE,* Vol. 68, May, pp. 593–619.

Schilling, D. L., L. B. Milstein, R. L. Pickholtz, and R. W. Brown. 1980. "Optimization of the Processing Gain of an *M*-ary Direct Sequence Spread Spectrum Communication System." *IEEE Trans. Commun.,* Vol. COM-28, Aug., pp. 1389–1398.

Schilling, D. L,, R. L. Pickholtz, and L. B. Milstein, guest eds. 1990. "Spread Spectrum Communications I." Special issue, *IEEE J. Sel. Areas Commun.,* Vol. SAC-8, May.

Scholtz, R. A. 1982. "The Origins of Spread-Spectrum Communications." *IEEE Trans. Commun.,* Vol. COM-30, May, pp. 822–854.

Schwartz, L. 1950. *Theorie des Distributions,* Vol. 1. Paris: Hermann.

Schwartz, M. 1987. Telecommunications Networks: Protocols, Modeling, and Analysis. Reading, Mass.: Addison-Wesley.

Schwartz, M. 1990. *Information Transmission, Modulation, and Noise,* 4th ed. New York: McGraw-Hill.

Shannon, C. E. 1948. "A Mathematical Theory of Communication." *Bell Syst. Tech. J.,* Vol. 27, July, pp. 379–423; Oct., pp. 623–656.

Shannon, C. E. 1959. "Coding Theorems for a Discrete Source with a Fidelity Criterion." *IRE Natl. Conv. Rec.,* Pt. 4, Mar., pp. 142–163.

Shannon, C. E., and W. Weaver. 1949. *The Mathematical Theory of Communication.* Urbana, Illinois.: University of Illinois Press.

Simon, M. K., J. K. Omura, R. A. Scholtz, and B. K. Levitt. 1985. *Spread Spectrum Communications,* Vols. I and II, Rockville, Md.: Computer Science Press.

Sklar, B. 1988. *Digital Communications: Fundamentals and Applications.* Englewood Cliffs, N.J.: Prentice Hall.

Spragins, J. D., J. L. Hammond, and K. Pawlikowski. 1991. *Telecommunications: Protocols and Design.* Reading, Mass.: Addison-Wesley.

Stallings, W. 1984. "Local Network Performance." *IEEE Commun. Mag.,* Vol. 22, Feb., pp. 27–36.

Sunde, E. D. 1961. "Pulse Transmission by AM, FM, and PM in the Presence of Phase Distortion." *Bell Syst. Tech. J.,* Vol. 40, Mar., pp. 353–422.

Tanenbaum, A. S. 1988. *Computer Networks.* Englewood Cliffs, N.J.: Prentice Hall.

Taub, H., and D. L. Schilling. 1986. *Principles of Communication Systems.* New York: McGraw-Hill, 2nd ed.

Temple, G. 1953. "Theories and Applications of Generalized Functions." *J. London Math. Soc.,* Vol. 28, pp. 134-148.

Thapar, H. K. 1984. "Real-Time Application of Trellis Coding to High-Speed Voiceband Data Transmission." *IEEE J. Sel. Areas Commun.,* Vol. SAC-2, Sept., pp. 648–658.

Thomas, G. B., Jr. 1968. *Calculus and Analytic Geometry.* Reading, Mass.: Addison-Wesley.

Ungerboeck, G. 1982. "Channel Coding with Multilevel/Phase Signals." *IEEE Trans. Inf. Theory,* Vol. IT-28, Jan., pp. 55–67.

Viswanathan, R., and K. Taghizadeh. 1988. "Diversity Combining in FH/BFSK Systems to Combat Partial Band Jamming." *IEEE Trans. Commun.,* Vol. COM-36, Sept., pp. 1062–1069.

Viterbi, A. J., and I. M. Jacobs. 1975. "Advances in Coding and Modulation for Noncoherent Channels Affected by Fading, Partial Band, and Multiple Access Interference." In *Advances in Communication Systems,* Vol. 4. A. V. Balakrishnan, ed. New York: Academic Press.

Wozencraft, J. M., and I. M. Jacobs. 1965. *Principles of Communication Engineering.* New York: Wiley.

Wyner, A. D. 1981. "Fundamental Limits in Information Theory." *Proc. IEEE,* Vol. 69, Feb., pp. 239–251.

Ziemer, R. E., and R. L. Peterson. 1985. *Digital Communications and Spread Spectrum Systems.* New York: Macmillan.